REVISE AQA GCSE (9–1)

Biology

REVISION GUIDE

Foundation

Series Consultant: Harry Smith

Author: Nigel Saunders

Contents

A small bit of small print:
AQA publishes Sample Assessment Material and the Specification on its website. This is the official content and this book should be used in conjunction with it. The questions in Now try this have been written to help you practise every topic in the book. Remember: the real exam questions may not look like this.

Microscopes and magnification

Elements

The **light microscope** was invented about 350 years ago. It uses light to form magnified images of cells and other tiny objects.

The best light microscope can produce magnifications of about 2000 times: ×2000.

The **electron microscope** was invented in the last century. It uses electrons to produce magnifications of up to 2 million times.

The higher magnification and resolving power of electron microscopes allow scientists to see and understand sub-cellular structures.

Resolving power

A microscope with a high **resolution** allows you to distinguish between two objects that are very close together.

Microscope	Resolution (nm)
light microscope	200
electron microscope (scanning)	10
electron microscope (transmission)	0.1

Maths skills Magnification

You can calculate **magnification** using:

$$\text{magnification} = \frac{\text{size of image}}{\text{size of real object}}$$

You can rearrange the equation to find the size of the real object, if you know the magnification and the size of the image:

$$\text{size of real object} = \frac{\text{size of image}}{\text{magnification}}$$

Worked example

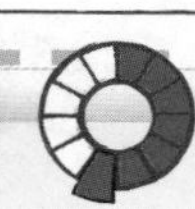

A light microscope has a ×10 eyepiece lens and a ×40 objective lens. The image of an onion cell viewed using this microscope is 0.2 mm long.

(a) Calculate the total magnification of the microscope. **(1 mark)**

total magnification = 10 × 40 = ×400

(b) Use your answer to part (a) to calculate the real length of the cell. **(1 mark)**

$$\frac{\text{size of image}}{\text{magnification}} = \frac{0.2\text{ mm}}{400} = 0.0005\text{ mm}$$

1 mm = 1000 µm

0.0005 × 1000 = 0.5 µm

Scale bars

Diagrams and photos of microscope images often have a scale bar on them.

A scale bar lets you:

- easily see the real sizes of the different objects in the diagram or photo
- calculate the magnification, if you do not know what this is.

If you want to calculate the magnification, you need to use a ruler to measure the real length of the scale bar.

It is better to make your measurements in millimetres, mm (rather than in centimetres, cm).

Worked example

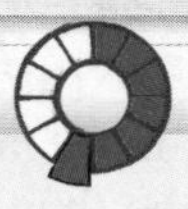

The diagram shows a cell with a scale bar.

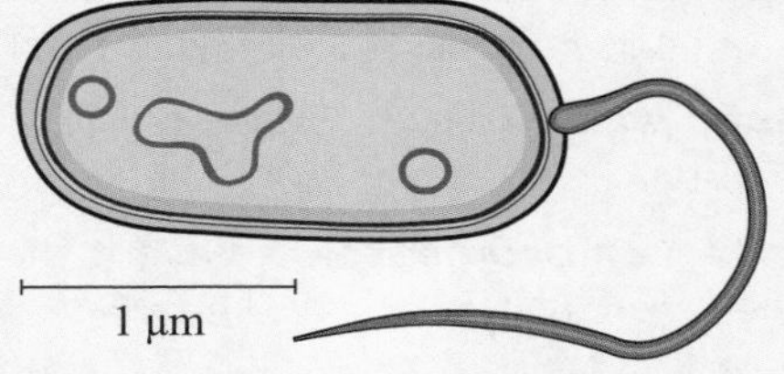

Calculate the magnification of this image. **(2 marks)**

length of scale bar on diagram = 20 mm

= 20 × 1000 = 20 000 µm

real length (from scale bar) = 1 µm

$$\text{magnification} = \frac{20\,000}{1} = \times 20\,000$$

Now try this

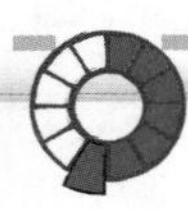

A light microscope has a ×5 eyepiece lens and a ×20 objective lens. The image of a cell is 15 mm long. Calculate the actual length of the cell. **(2 marks)**

Animal and plant cells

Animal **cells** and plant cells contain smaller structures which have different functions.

Structures of animal cells and plant cells

Generalised animal cell

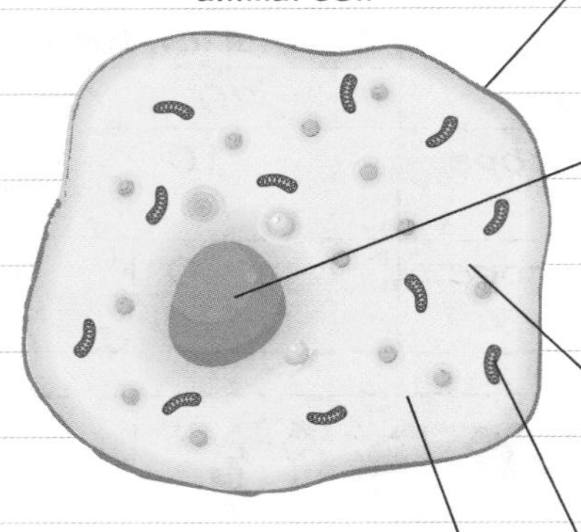

cell membrane: controls what enters and leaves the cell, e.g. oxygen, carbon dioxide, glucose

nucleus: a large structure that contains genes that control the activities of the cell

cytoplasm: jelly-like substance that fills the cell – many reactions take place here

mitochondria (single: mitochondrion): tiny structures where respiration takes place, releasing energy for cell processes

ribosomes (present in the cytoplasm but not visible at this size): where proteins are made (protein synthesis)

Generalised plant cell

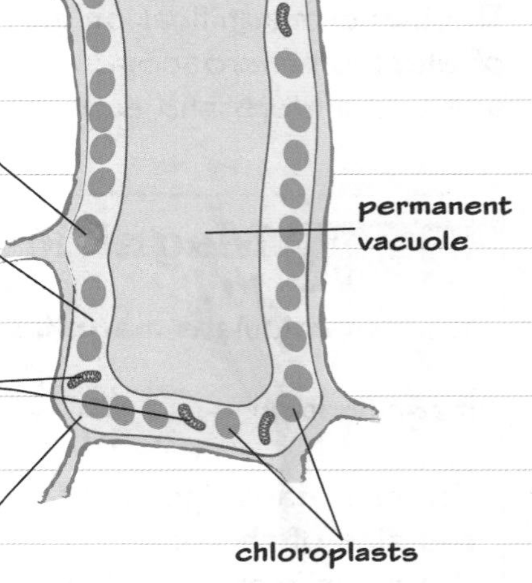

cell wall

permanent vacuole

chloroplasts

Worked example

Name the three structures that are present in most plant and algal cells, but are not present in animal cells. Describe the function of each structure. **(3 marks)**

Plant and algal cells contain a permanent **vacuole**. This is filled with cell sap and helps to keep the cell rigid.

They have a **cell wall** made of cellulose, which strengthens the cell.

They also contain **chloroplasts**. These are the structures where photosynthesis takes place to make food for the cell.

Algae are simple aquatic organisms. Their cells have a similar structure to plant cells, so algae used to be classified as plants. They are now classified as protists (see pages 36 and 84).

Take care not to confuse the cell wall with the cell membrane.

Bacterial cells also have cell walls, but these are not made from cellulose.

Now try this

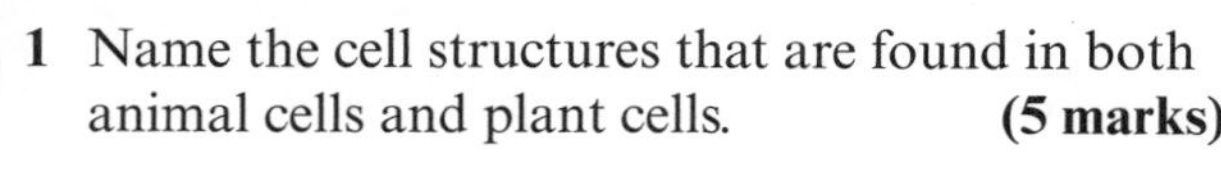

1 Name the cell structures that are found in both animal cells and plant cells. **(5 marks)**

2 Muscle cells contain more mitochondria than skin cells. Suggest an explanation for this. **(2 marks)**

3 Cells found in plant roots do not usually contain chloroplasts. Suggest an explanation for this. **(2 marks)**

Eukaryotes and prokaryotes

Animal cells and plant cells are **eukaryotic** cells, but bacterial cells are **prokaryotic** cells.

Bacterial cells

Eukaryotic cells and prokaryotic cells have:
- a cell membrane
- cytoplasm.

But prokaryotic cells:
- are surrounded by a cell wall
- do not have a nucleus.

A single loop of **chromosomal DNA** lies free in the cytoplasm. This carries most of the bacterial genes.

cell membrane

Some bacteria have a **flagellum** to help them move.

Ribosomes are tiny structures that make proteins.

Some bacteria have extra circles of DNA called **plasmid DNA**. Plasmids contain additional genes that are not found in chromosomes.

Many bacteria have a **cell wall** for protection, but it is made of different substances to plant cell walls.

Standard form

Numbers in standard form have two parts.

$$7.3 \times 10^{-6}$$

This part is a number greater than or equal to 1 and less than 10.

This part is a power of 10.

You can use standard form to write very large or very small numbers.

$$920\,000 = 9.2 \times 10^{5}$$

Numbers greater than 10 have a positive power of 10.

$$0.007\,03 = 7.03 \times 10^{-3}$$

Numbers less than 1 have a negative power of 10.

Maths skills: Estimation

You can **estimate** an answer to a calculation. To do this, you round all the values involved to 1 significant figure.

For example, you can estimate the area of a 125 µm × 38 µm plant cell.

125 µm is 100 µm to 1 significant figure, and 38 µm is 40 µm to 1 significant figure.

The estimated area = 100 µm × 40 µm = 4000 μm^2

Estimates are useful if you need a rough idea of a value, or you want to check an answer. In this example, the estimated area is similar to the calculated area (4750 μm^2).

Worked example

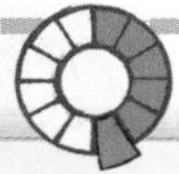

An animal cell is 0.01 mm long and a bacterial cell is 1 µm long.

(a) Give these lengths in metres, m, in standard form. **(2 marks)**

$0.01\text{ mm} = 0.01 \times 10^{-3}\text{ m} = 1 \times 10^{-5}\text{ m}$

$1\ \mu m = 1 \times 10^{-6}\text{ m}$

(b) How many orders of magnitude greater is the size of the animal cell? **(1 mark)**

0.01 mm = 0.01 × 103 µm = 10 µm

So the animal cell is $\frac{10}{1}$ = 10 times larger. This is one order of magnitude.

These units use the prefixes milli, m, and micro, µ. They refer to powers of 10:

milli- 10^{-3} | micro- 10^{-6} | nano- 10^{-9}

1 nm (nanometre) is 1×10^{-9} m.

You can also work this out using the powers of 10 in the lengths given in standard form.

The difference between 5 and 6 is 1, showing one order of magnitude difference.

Now try this

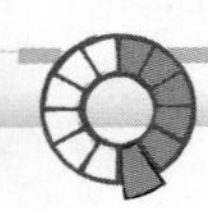

(a) Give these lengths in metres, m, in standard form:
(i) 75 µm (the diameter of a plant cell); (ii) 750 nm (the diameter of a mitochondrion). **(2 marks)**

(b) Determine how many orders of magnitude greater a plant cell is than a mitochondrion. **(1 mark)**

Specialised animal cells

Animal cells may be specialised to carry out a particular function.

Ciliated cells

Cilia are tiny hair-like structures on the surface of some cells. They can move in organised ways to push substances past the surface of the cell.

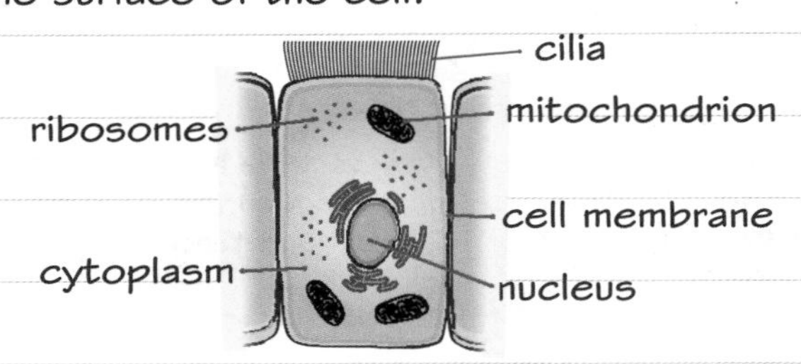

Ciliated cells line the **trachea** or windpipe. These sweep mucus away from the lungs, carrying away dirt and bacteria.

Differentiation

As an organism develops, cells **differentiate** to form different specialised cells:

- ✓ Most types of animal cell differentiate at an early stage.
- ✓ Cells acquire different structures so they can carry out a particular function.
- ✓ Cell division in mature animals is mainly for repair and replacement.

You can revise the structures in generalised animal cells on page 2 and stem cells on page 10.

Sperm cells

Sperm cells carry the male parent's genetic information to the female parent's egg cell.

They may have to travel a relatively long way, then get through the surface of the egg cell.

acrosome

nucleus

tail

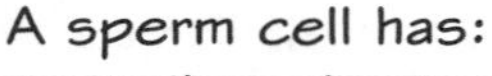

A sperm cell has:

- a tail, and many mitochondria to release the energy needed for the tail to move
- enzymes in its acrosome to digest the outer layers of an egg cell.

Nerve cells

Nerve cells carry electrical impulses from one part of the body to another, which may be a long way apart. Nerve cells often connect with one another.

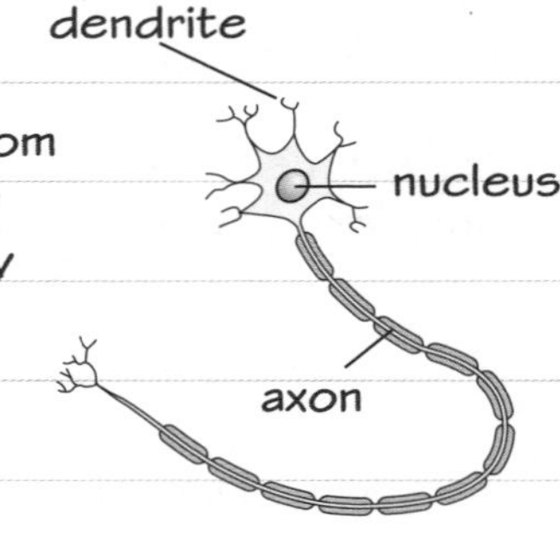

A typical nerve cell has:

- a long axon to connect distant parts
- tiny finger-like dendrites that make connections with other nerve cells.

Worked example

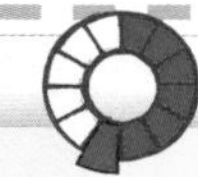

The diagram shows muscle cells. Describe, using information shown, how they are specialised for their function. **(2 marks)**

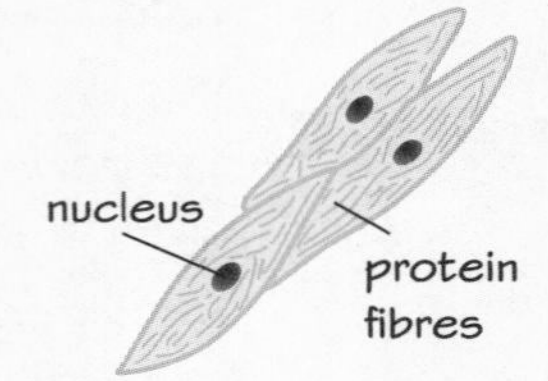

The protein fibres slide over each other, making the muscle cells contract.

Muscle cells also contain many mitochondria. These supply the energy needed to keep the muscles working.

Now try this

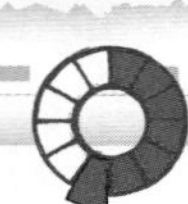

Different specialised cells have different shapes. Explain how the shapes of ciliated cells in the trachea and sperm cells are related to their functions. **(4 marks)**

Specialised plant cells

Plant cells may be **specialised** to carry out a particular function.

Root hair cells

Root hair cells are found on the surface of plant roots. They absorb water and mineral ions.

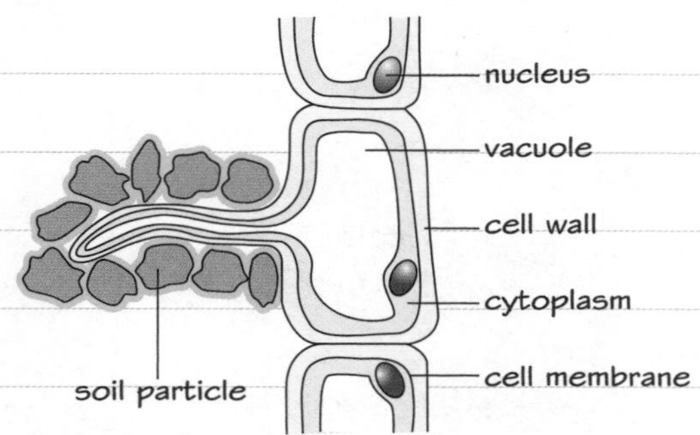

A root hair cell has:

- a large vacuole to increase the rate of absorption of water by **osmosis**
- many mitochondria to release the energy needed for the **active transport** of mineral ions.

Differentiation

As an organism develops, cells **differentiate** to form different specialised cells:

- ✓ Meristem cells in the tips of roots and shoots are undifferentiated cells.
- ✓ Cells acquire different structures so they can carry out a particular function.

Unlike most types of animal cells, many types of plant cells keep the ability to differentiate through their life.

You can revise the structures in generalised plant cells on page 2 and stem cells on page 10.

You can revise diffusion on page 11, osmosis on page 13 and active transport on page 15.

Xylem cells

Xylem cells form the **transport tissue** that carries water and mineral ions from roots to the rest of the plant. Lignin builds up in xylem cells. These die and form hollow tubes.

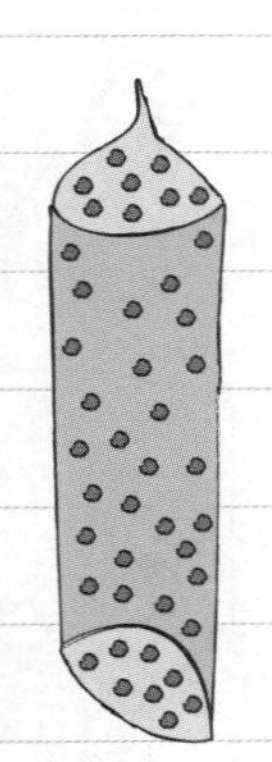

Xylem cells have:

- **lignin** to provide strength and support
- no end walls, so water and mineral ions can flow easily through the plant
- **pits** that let water and ions move out.

Worked example

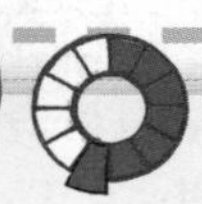

Phloem cells form the transport tissue that carries dissolved sugars through a plant.

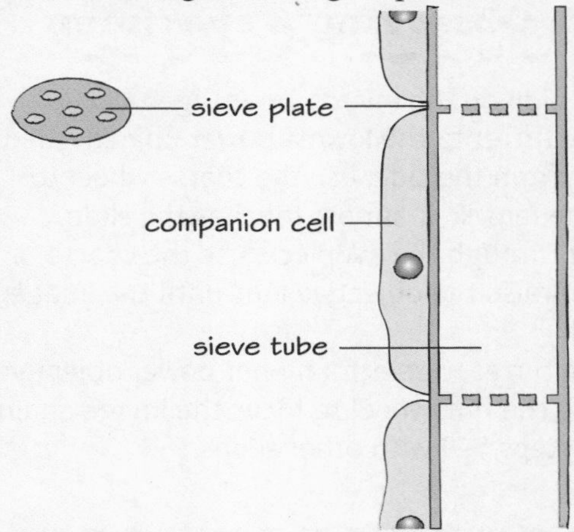

Explain **two** ways in which phloem cells are specialised for their function. **(4 marks)**

The cell walls between cells form sieve plates, which let solutions move from cell to cell. They have companion cells which contain lots of mitochondria. These release the energy needed for the active transport of sugars.

Now try this

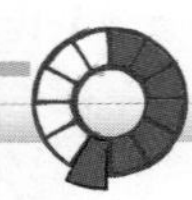

Root hair cells have large surface areas to increase the rate of absorption of water and mineral ions. Describe the structure that provides this feature. **(2 marks)**

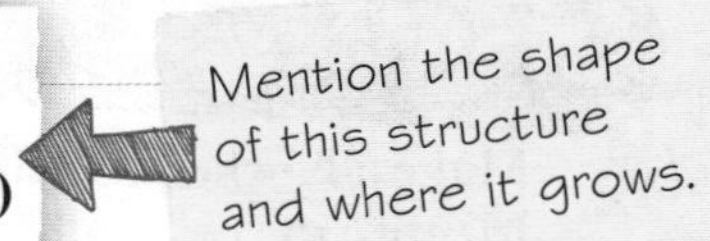

Using a light microscope

Practical skills

You can use a **light microscope** to observe, draw and label a selection of plant and animal cells.

Core practical

Observing cells

Aim

to make a slide with onion cells, then use a microscope to observe, draw and label these and other cells

Apparatus

- small piece of onion
- forceps
- microscope slide with coverslip
- iodine solution
- prepared slides of animal and plant cells
- light microscope

Onion skin is easily peeled away in single layers of cells.

Wear eye protection when you handle the iodine solution. Wash off spills straight away because it will stain skin and clothes.

Method: Preparing a specimen

1. Use the forceps to peel a thin layer of epidermal tissue from the onion.
2. Carefully lay the thin layer on the slide.
3. Add one or two drops of iodine solution.
4. Lower the coverslip over the onion layer. Remove excess liquid with tissue paper.

You may find it easier to position the layer of onion if you put a drop of water on the slide first. Don't let the tissue curl over on itself.

Place one edge of the coverslip on the slide at 45°. Gently lower the other edge so that liquid and bubbles are pushed out.

Method: Observing a specimen

5. Place a slide on the microscope stage.
6. Turn the turret to the lowest power objective lens.
7. Looking from the side, use the coarse wheel to lower the lens so it almost touches the slide.
8. Looking through the eyepiece, use the coarse wheel to raise the objective lens until the image is focused.
9. Turn the turret to select a higher power objective lens. Use the fine wheel to focus the image again.
10. Repeat steps 5–9 with other slides.

Results

Make a pencil drawing of one or more cells:

- Draw outlines of what you see.
- Use shading only to distinguish between structures.
- Identify structures using labels with straight lines.
- Write the magnification on your drawing.

You can revise calculating magnification on page 1.

The light microscope

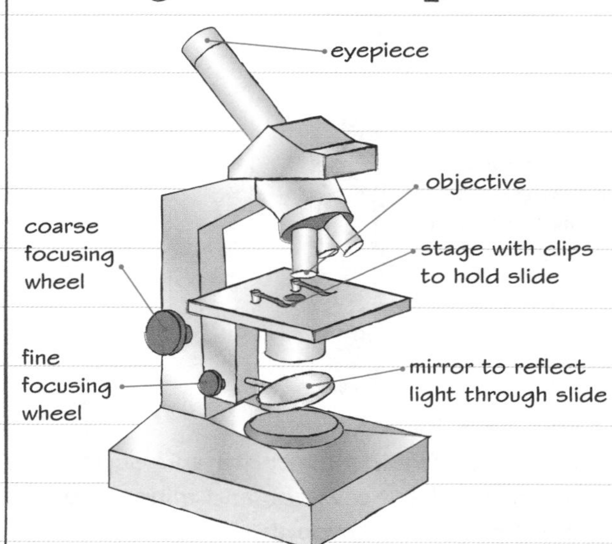

Some modern microscopes have a lamp rather than a mirror.

Now try this

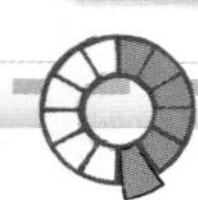

Make a labelled drawing of the onion epidermal cells.
Identify the cell wall, nucleus, cytoplasm and a chloroplast. **(5 marks)**

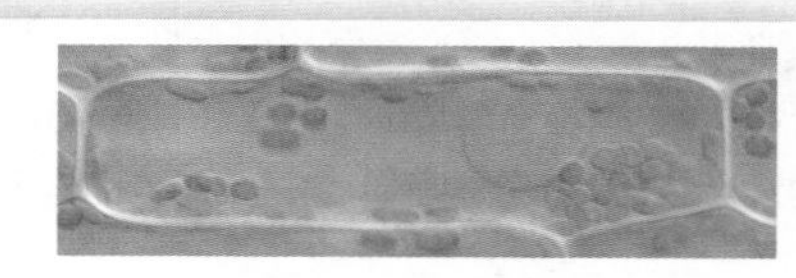

Aseptic techniques

Bacteria can be grown in a nutrient broth solution or as colonies on an agar gel plate.

Sterilising apparatus and media

Killing microorganisms is called **sterilisation**.

For use in the lab:

- **sterile** Petri dishes, irradiated with gamma radiation at the factory, can be used
- culture media (such as nutrient broth or agar gel) can be sterilised using an **autoclave**, which uses steam at high temperature under pressure
- apparatus can be kept sterile during working by using a flame.

Binary fission

Bacteria reproduce by simple cell division, called **binary fission**.

Bacteria can divide as often as once every 20 minutes if:

- they have enough **nutrients**, and
- they are at a suitable **temperature**.

When investigating the growth of bacteria, it is important to use uncontaminated cultures, and sterile apparatus and culture media.

Sterilising apparatus

A wire loop is used to transfer a sample of a bacterial culture from one place to another.

The loop is sterilised in a hot flame and then cooled, before using it to transfer microorganisms to the culture medium.

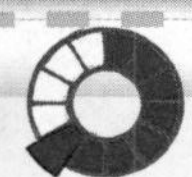

Human body temperature is 37 °C. In school laboratories, cultures should generally be incubated at 25 °C. This is to reduce the risk of harmful bacteria growing.

Incubating Petri dishes

The layer of sterile agar gel on a Petri dish is inoculated with bacteria using a wire loop:

- The Petri dish is opened upside down near a flame.
- The loop is swept over the agar gel.
- The lid is replaced and sealed.
- The dish is labelled with the worker's name, the date, and what it contains.

The dish is not sealed all round. This is to stop the growth of harmful anaerobic bacteria.

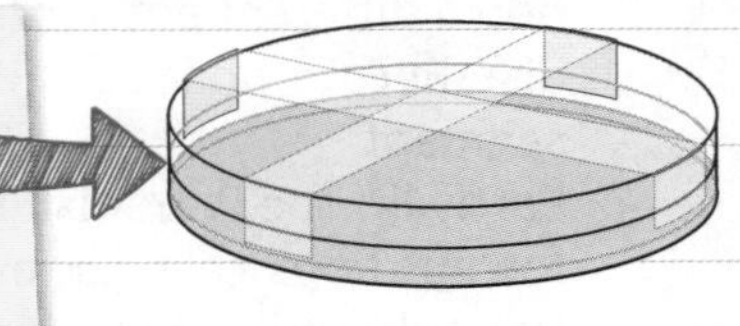

Worked example

The mean division time for a strain of bacteria is 0.5 hours. A culture with 1000 of these bacteria is incubated. Calculate the number of bacteria after 4 hours. **(3 marks)**

number of divisions $= \frac{4}{0.5} = 8$

number of bacteria $= 1000 \times 2^8$

$= 1000 \times 256$

$= 256000$

Maths skills: Using powers

Calculators usually have a button marked x^y

To calculate 2^8 (two to the power eight) on your calculator, enter: 2 x^y 8 =

Some calculators have a button marked y^x

If neither button is present, instead enter:

$2 \times 2 \times 2 \times 2 \times 2 \times 2 \times 2 \times 2$

Now try this

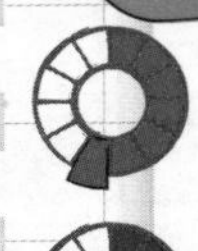

1 Describe **three** ways in which apparatus and culture media can be sterilised. **(3 marks)**

2 Suggest an explanation for why the laboratory bench is wiped with an antiseptic solution before working with bacteria. **(2 marks)**

3 The mean division time for a strain of bacteria is 40 minutes. A culture of 2500 of these bacteria is incubated. Calculate the number of bacteria after 4 hours. **(2 marks)**

An antiseptic is a substance that kills microorganisms or prevents their growth.

Investigating microbial cultures

Practical skills You can measure the effect of antiseptics and antibiotics by measuring **zones of inhibition.**

Core practical

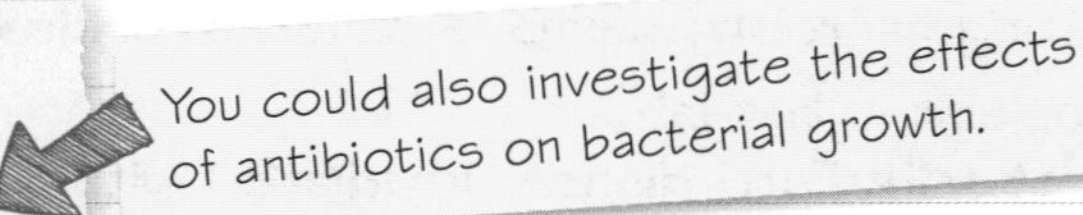

The effects of antiseptics

Aim

to investigate the effect of antiseptics on bacterial growth using agar plates

Apparatus

Petri dish with bacteria spread over nutrient agar • filter paper discs, about 4 mm in diameter • three different antiseptics • forceps • adhesive tape • permanent marker pen • Bunsen burner and heat-resistant mat

Your teacher or school technician will prepare the Petri dish for you. They will:

- add a few drops of a bacterial culture to the agar on the Petri dish
- use a sterile glass spreader to spread the bacteria over the agar.

Certain strains of *E. coli* are suitable to use.

Method

Make sure you wipe the bench with a disinfectant solution, then dry it with clean paper towels before starting.

1 Turn the Petri dish upside down and mark the bottom into three equal segments.
2 Mark a spot in the middle of each segment. Number each segment. Write your name and the date on the bottom.
3 Add a different antiseptic to each disc. Record what you added to each segment.
4 Partly lift the Petri dish and use the forceps to place a disc over each spot.
5 Use adhesive tape to secure the lid.
6 Incubate the plate for 2 days at 25 °C.

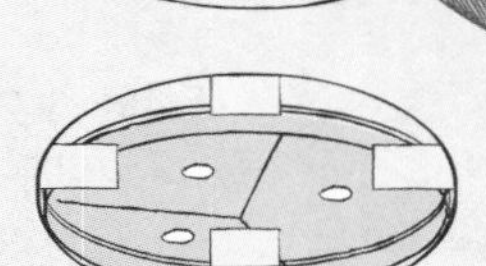

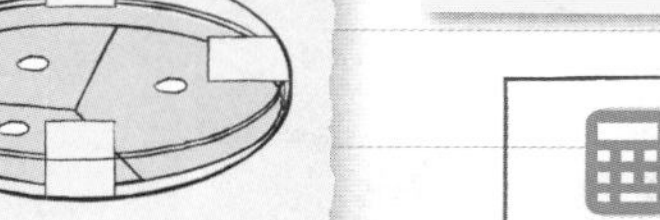

Before you use the forceps, briefly pass the end of the forceps through a Bunsen burner flame to sterilise them.

Results

Measure the diameter of the clear zone around each paper disc:

- Measure in millimetres, mm.
- Make two measurements at 90° to each other, and calculate the mean diameter.
- Calculate the area of each clear zone, giving your answers in millimetres, mm^2.

Maths skills The area of a circle

You can calculate the area of a circle using:

$$\text{area} = \pi r^2$$

where r = radius.

Divide the diameter by 2 to find the radius.

Now try this

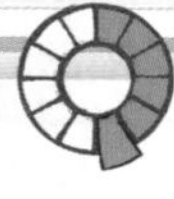

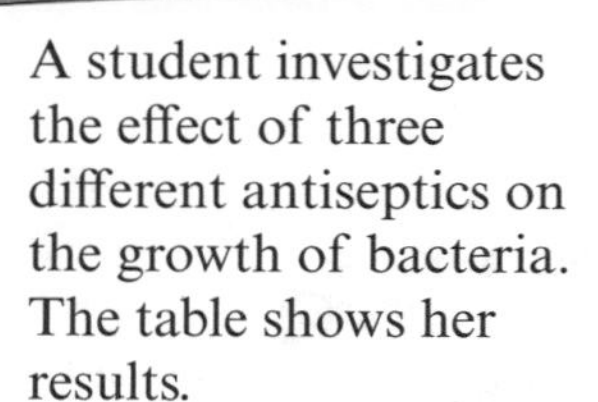

A student investigates the effect of three different antiseptics on the growth of bacteria. The table shows her results.

	Diameter of the clear zone (mm)			
Antiseptic	Measurement 1	Measurement 2	Mean value	Area (mm^2)
A	14	16	15	177
B	24	26		
C	19	21		

(a) Calculate the mean diameter and mean area for antiseptics B and C. **(4 marks)**

(b) Explain what the results show. **(2 marks)**

Mitosis

The nucleus of a cell contains **chromosomes** made of **DNA** molecules which carry a large number of **genes**. The chromosomes are normally found in pairs in body cells.

The cell cycle

There are three main stages in the **cell cycle**:

1. **interphase**, the stage where the cell:
   - grows
   - increases the numbers of structures such as ribosomes and mitochondria
   - **replicates** its DNA to form two copies of each chromosome.
2. **mitosis**, the stage where:
   - one set of chromosomes is pulled to each end of the cell
   - the nucleus divides.
3. **cell division**, the stage immediately after mitosis where:
   - cell membranes and cytoplasm divide
   - two identical cells form.

Interphase

A cell spends most of its life in interphase.

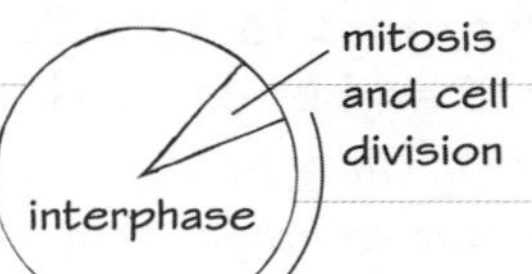

Three things to remember

- mi-**T**o-sis makes **T**wo cells
- m**I**-tosis makes cells that are genetically **I**dentical to the parent cell
- **D**iploid means **D**ouble (two sets of) chromosomes.

Parent cell and two daughter cells

The cell that divides is called the **parent cell**. → The parent cell divides to form two **daughter cells**.

The parent cell is a **diploid** cell. This means it has two sets of chromosomes.

Before the parent cell divides, each chromosome is copied exactly.

When the cell divides in two, each cell gets one copy of each chromosome.

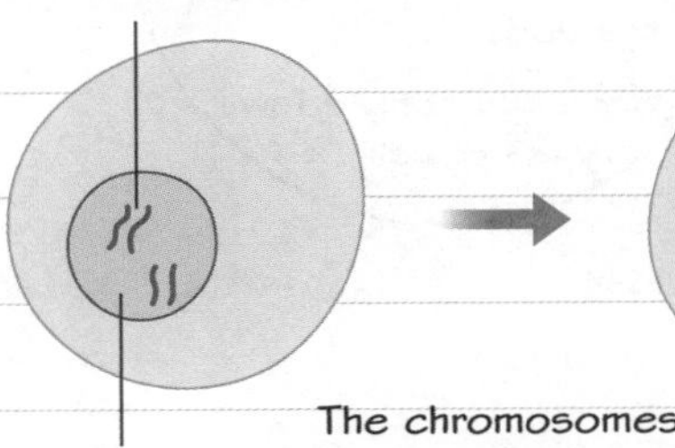

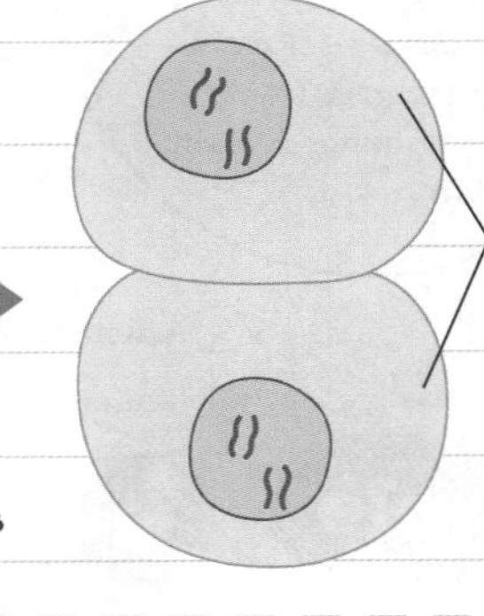

The daughter cells are genetically identical to each other and the parent cells. They are also diploid cells.

The chromosomes are drawn short here, and coloured, so it is easier to see what is happening. They don't really look like this.

Worked example

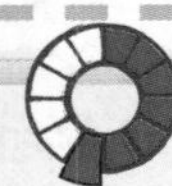

Give **three** situations in which mitosis is used to produce new cells. **(3 marks)**

Mitosis produces new body cells for:

- growth
- repair to damaged parts of the body.

It also produces new individuals by asexual reproduction.

Cell division by mitosis is important in the growth and development of multicellular organisms.

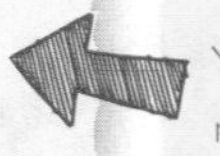

You can revise asexual reproduction on page 68.

Now try this

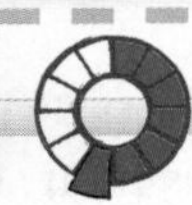

Explain why the daughter cells produced by mitosis have the same number of chromosomes as the parent cell. **(2 marks)**

Stem cells

Stem cells are undifferentiated cells in an organism. They can produce many more cells of the same type, and specialised cells by **differentiation**.

Stem cells in animals

A zygote is a single cell formed when an egg cell and sperm cell join together. This divides to form a ball of cells called an **embryo**:

- the embryo contains **embryonic stem cells**, which differentiate to form many different specialised cell types.

Adult animals retain some stem cells. The bone marrow contains **adult stem cells**. These form many cell types, including blood cells.

Stem cells in plants

The **meristems** in plants contain stem cells.

Meristem tissue is found at the tips of roots and shoots. It differentiates throughout the life of the plant:

- in roots to form root hair cells, xylem, phloem and other cells in the root
- in shoots to form mesophyll cells, xylem, phloem and other cells in the stems and leaves.

Stem cell treatment

Stem cells may be able to help to treat conditions such as diabetes and paralysis. Injected stem cells may be able to replace damaged or missing cells in the pancreas or nervous system.

✓ – Advantage ✗ – Disadvantage ! – Risk

Worked example

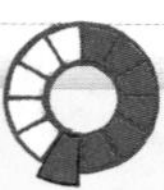

Stem cells from meristems in plants can be used to produce clones rapidly and economically. Explain one possible use of producing plants this way. **(2 marks)**

If a crop plant has a useful feature, such as resistance to disease, it could be cloned. This would provide large numbers of these plants for farmers to grow.

Clones are genetically identical individuals.

Cloning could also save rare plant species from extinction by producing large numbers of these plants.

You can revise cloning in plants on page 77.

Now try this

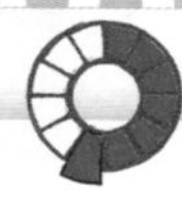

Therapeutic cloning is a way to produce an embryo using a cell from a patient. It involves replacing the nucleus in an egg cell with the nucleus from one of the patient's cells. The egg cell is then stimulated to make it divide to form an embryo.

Explain an advantage and a disadvantage of treating a patient with stem cells produced in this way. **(4 marks)**

Diffusion

Substances may move into and out of cells by **diffusion** across cell membranes.

Net movement of particles

Diffusion is the spreading out of:
- the particles of a gas, or
- the particles of any substance in solution.

Diffusion results in an overall movement of particles down a **concentration gradient**:
- from an area of higher concentration to an area of lower concentration.

The rate of diffusion increases when:
- the difference in concentration increases, so the concentration gradient increases
- the temperature increases, so the particles move more quickly.

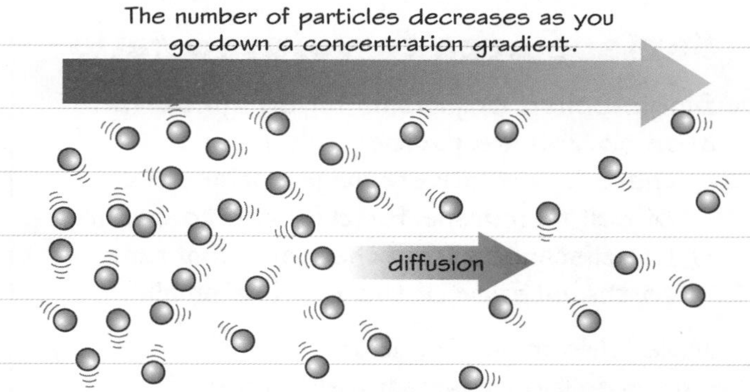

Diffusion explains why you can smell cooking from the next room, and why ink spreads through a glass of water without stirring.

Diffusion across membranes

Diffusion also happens across membranes.

In addition to the two factors above, the rate of diffusion increases when:
- the **surface area** of the membrane **increases**
- the **thickness** of the membrane **decreases**.

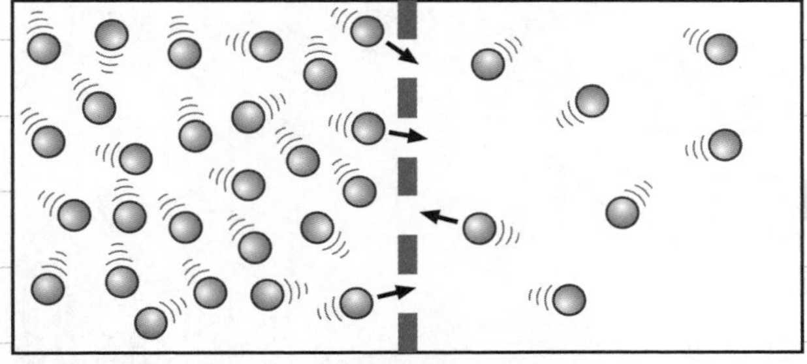

Molecules pass through the membrane in both directions. The rate of movement from left to right is greater than the rate of movement from right to left.

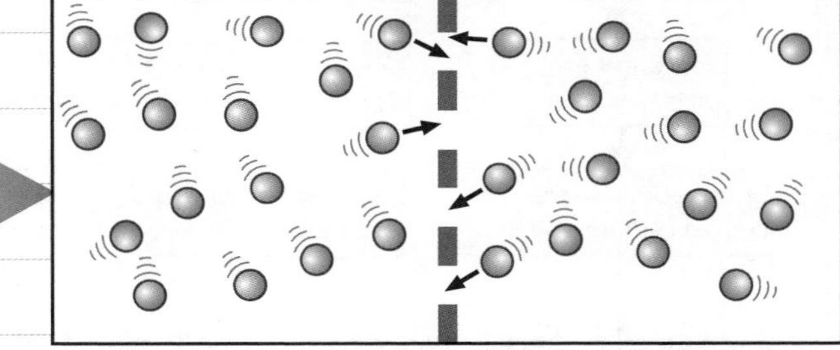

There is a net movement of particles from left to right until their concentration is the same on both sides.

Particles remain in random and continuous motion before, during and after diffusion.

Worked example

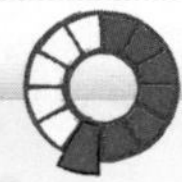

Describe **three** examples of the movement of substances into or out of animal cells by diffusion. **(3 marks)**

During gas exchange, oxygen moves into cells by diffusion, and carbon dioxide moves out of the cells by diffusion.

Urea is a waste product. It moves out of cells into the blood plasma by diffusion.

Diffusion also happens in plants, such as gas exchange in the leaves during photosynthesis.

You can revise gas exchange in the lungs on page 24, and the function of the kidneys on page 60.

Now try this

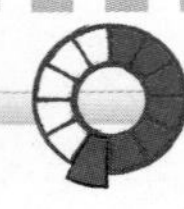

1 State what happens to the rate of diffusion as the temperature increases. **(1 mark)**
2 Describe what is meant by diffusion. **(2 marks)**

Exchange surfaces

Multicellular organisms have adaptations to allow effective exchange of materials.

Surface area to volume ratio

Single-celled organisms have high **surface area to volume ratios**. This means:

- there is a relatively large surface area of cell membrane for diffusion to occur
- the distances from the cell membrane to the interior of the cell are small.

Molecules move in and out of a single-celled organism well enough to meet its needs.

However, multicellular organisms need:

- **exchange surfaces** to provide large surface area to volume ratios
- **transport systems** to maintain high concentration gradients.

Maths skills: Calculating surface area to volume ratio

Imagine a cube-shaped cell with sides 1 μm long. It has six equal square sides:

- surface area = $6 \times (1 \times 1) = 6\ \mu m^2$
- volume = $1 \times 1 \times 1 = 1\ \mu m^3$

The cell's surface area to volume ratio is: $\frac{6}{1} = 6$

Surface area to volume ratio decreases with increasing size. For example, for cubes:

Length of side (μm)	Surface area to volume ratio
10	0.6
100	0.06

Increasing the effectiveness of an exchange surface

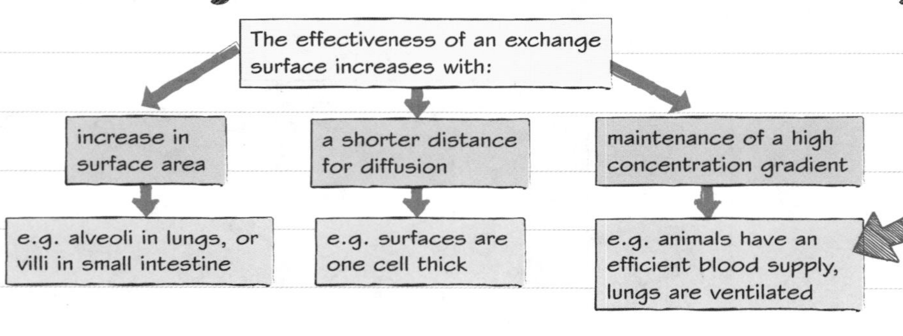

The fast removal of substances maintains a steep concentration gradient so diffusion is faster.

For example, the lining of the small intestine is folded into many tiny finger-like structures called villi. These provide:

- a large surface area for diffusion
- a single layer of cells for a short diffusion path
- a network of capillaries to carry food molecules away.

Worked example

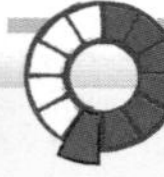

The diagram shows the gills in a fish. Describe how they are adapted for gas exchange. **(2 marks)**

The gills have tiny finger-like gill filaments. These greatly increase the surface area available for gas exchange.

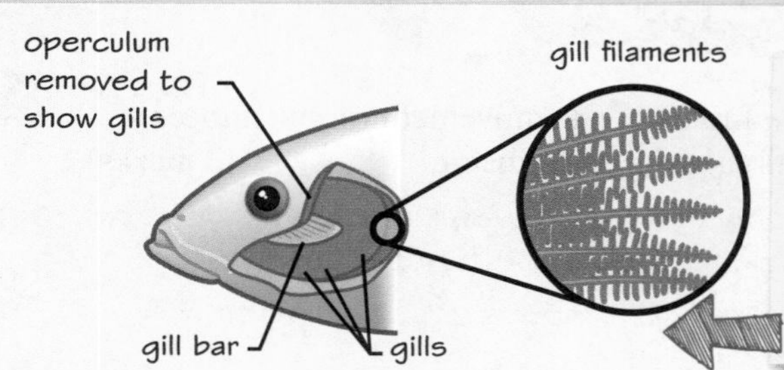

You can revise the adaptations of the lungs on page 24, and of leaves on page 29.

Now try this

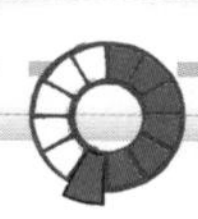

1 Describe **two** ways in which a high concentration gradient can be maintained. **(2 marks)**
2 Explain why multicellular organisms need exchange surfaces. **(2 marks)**

Osmosis

Water may move across cell membranes by **osmosis**.

Explaining osmosis

Osmosis is the diffusion of water:

- from a dilute solution
- to a concentrated solution
- through a **partially permeable membrane**.

A partially permeable membrane:

- allows small molecules such as water molecules to pass through it
- does not allow ions or larger molecules such as sucrose to pass through it.

Cell membranes are partially permeable. Water can diffuse across them but sugars cannot.

This diagram shows how there is a net movement of water from a dilute sucrose solution to a concentrated sucrose solution.

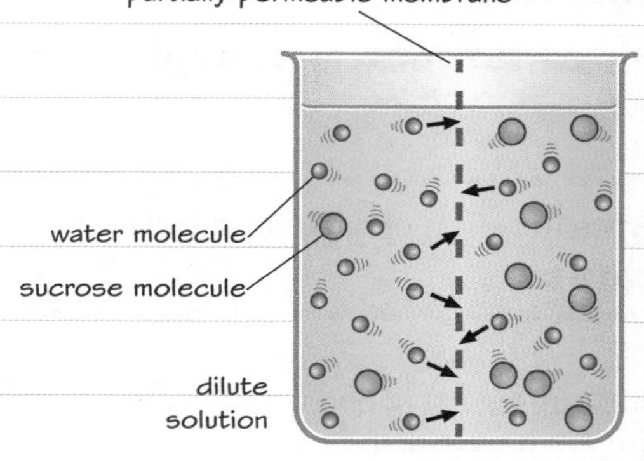

Investigating osmosis

Visking tubing is an artificial partially permeable membrane. It is supplied dry and is soaked in water to soften it before use.

The diagram on the right shows a typical osmosis experiment. The Visking tubing is:

- tied at the bottom with cotton thread
- filled with sucrose solution
- tightly fitted around a narrow glass capillary tube and tied with thread
- placed in a beaker of water.

There is a net movement of water from the beaker into the sucrose solution.

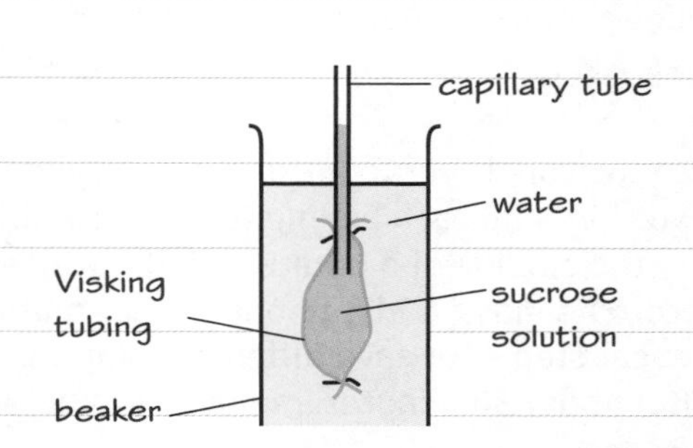

Over time, the volume of liquid inside the Visking tube increases. The level of liquid in the capillary tube rises.

Worked example

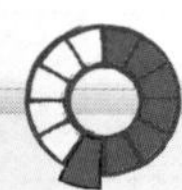

Red blood cells are placed on a microscope slide. A few drops of concentrated sucrose solution are added, and the red blood cells are observed using a microscope. They decrease in size. Explain these observations. **(2 marks)**

The sucrose solution is more concentrated than the cytoplasm in the red blood cells. Water moves across the cell membrane from the cytoplasm into the sucrose solution, causing the red blood cells to shrink.

You see similar results with other types of cell. Cells increase in size, and may even burst, if placed in water or very dilute solutions.

This is an example of osmosis, with the cell membrane acting as a partially permeable membrane.

You may see the term 'semi-permeable membrane' instead of 'partially permeable membrane'.

Now try this

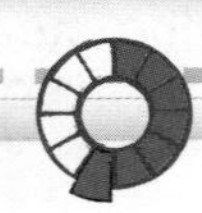

1 Describe what is meant by osmosis. **(3 marks)**

2 Red blood cells are placed on a microscope slide. Explain why the cells increase in size when a few drops of water are added. **(2 marks)**

Take another look at the Worked example.

Investigating osmosis

Practical skills

You can investigate the effect of **osmosis** on the mass of plant tissue.

Core practical

Concentration and osmosis

Aim

to investigate the effect of different sucrose concentrations on the mass of potato tissue

You could use different concentrations of sodium chloride instead.

Apparatus

- potato
- cork borer, cutting board and knife
- ruler with 1 mm scale divisions
- balance, precise to ±0.01 g
- different concentrations of sucrose
- boiling tubes with bungs and labels

A suitable range is 300 g/dm³ sucrose solution. Distilled water provides the 0 g/dm³ solution. If you use sodium chloride, use 25 g/dm³.

You could use labelled Petri dishes instead.

Method

1. Use the cork borer to make potato cylinders. Trim them to about 30 mm.
2. Two-thirds fill each boiling tube with a different sucrose solution.
3. Dry the outside of a potato cylinder with a paper towel, then measure and record its mass. Add it to one of the solutions.
4. Repeat step 3 for each different solution.
5. After about 30 minutes, remove each potato cylinder. Dry them and record their mass.

Make sure all the potato cylinders are the same length.

You could also measure and record the new lengths, and then analyse those results.

You could wait for up to 24 hours instead.

Results

Make a suitable table to record your results. This one has some example results in it.

Sucrose conc. (g/dm³)	Start mass (g)	End mass (g)	Change (g)	% change
0	2.00	2.53	+0.53	+26.5
75	2.15	2.28		
150	2.05	1.86		
225	1.96	1.58		
300	2.08	1.53	−0.55	−26.4

Analysis

1. Calculate the change in mass and percentage change in mass for each potato cylinder.
2. Plot a graph with:
   - percentage change in mass on the vertical axis
   - concentration of sucrose on the horizontal axis.

Maths skills

Calculating percentage change

percentage change in mass =

$$\frac{\text{(final mass − initial mass)}}{\text{initial mass}} \times 100$$

A negative value means the mass decreased.

Now try this

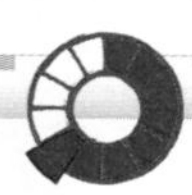

(a) Calculate the percentage change in mass for the results in the table. **(3 marks)**
(b) Plot a suitable graph using all the percentage changes in mass. **(3 marks)**
(c) Use your graph to estimate the concentration of the potato tissue. **(1 mark)**

This is when the percentage change in mass is zero.

Active transport

Unlike diffusion and osmosis, **active transport** requires **energy** from respiration to happen.

Explaining active transport

Active transport moves substances:

- from a more dilute solution
- to a more concentrated solution, and
- this requires energy from **respiration**.

Active transport lets substances move against a **concentration gradient**.

Cell membranes contain 'transport proteins'. The diagram shows how energy allows one of these molecules to change shape, which moves a solute particle into a cell.

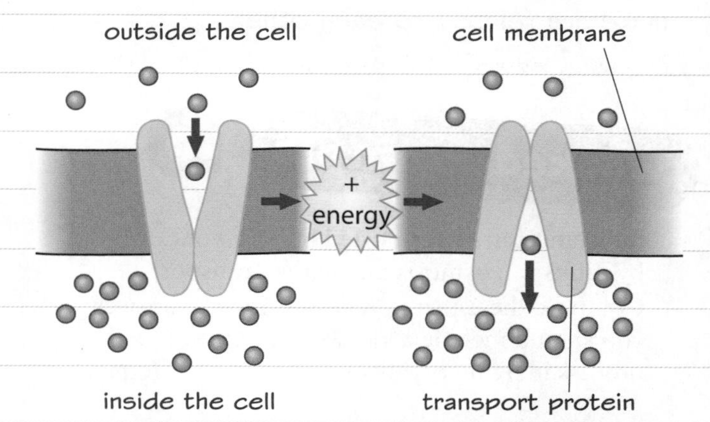

Active transport in the gut

Glucose is provided by the food we eat. It must be absorbed across the wall of the **small intestine** into the bloodstream. Active transport of glucose molecules means:

- glucose can be absorbed, even if its concentration in the blood is higher than its concentration in the small intestine.

It is important because cells need glucose for respiration. This happens in the mitochondria (you can revise cell structure on page 2).

Active transport in roots

Plants need **mineral ions** such as nitrate ions, NO_3^-, for healthy growth. The **root hair cells** absorb mineral ions from soil water. The concentration of ions in soil water is very low. Active transport of mineral ions means:

- plants can absorb mineral ions, even from very dilute solutions.

Transport processes

Process	Features
diffusion	• Substances move from high concentration of substance to low concentration. • Membranes are not needed. • No energy is needed.
osmosis	• Water moves from a dilute solution to a concentrated solution. • Partially permeable membrane is needed. • No energy is needed.
active transport	• Substances move from low concentration of substance to high concentration. • Membranes are needed. • Energy is needed.

Worked example

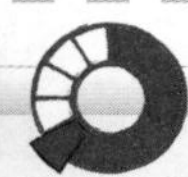

A farmer's field becomes flooded with water after heavy rain. Explain why the plants growing there are unable to absorb enough nitrate ions for healthy growth. **(4 marks)**

Plants absorb nitrate ions from the soil by active transport. This needs energy from cell respiration. Water-logged soil contains very little air. There is not sufficient oxygen for enough active transport by root hair cells.

Plants need nitrate ions so they can make proteins. Nitrate deficiency causes stunted growth in plants.

Some substances are toxic to cells because they prevent respiration. Active transport will also stop in the presence of these substances.

Now try this

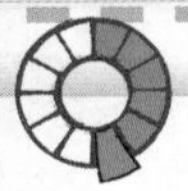

Give **three** differences between active transport and diffusion. **(3 marks)**

Extended response – Cell biology

There will be at least one 6-mark question on your exam paper. For these questions, you will need to think scientifically and structure your answer logically, showing how the points you make are related to each other.

For the questions on this page, you can revise cells and transport on pages 2 and 11–15.

Worked example

Substances are transported into and out of cells by diffusion, osmosis and active transport. Compare these three processes. In your answer, you should include a suitable example of each process in living organisms. **(6 marks)**

Diffusion is the spreading out of particles, producing a net movement from an area of higher concentration to an area of lower concentration. It is important for gas exchange in the lungs and gills.

Osmosis also involves diffusion, but only of water. It is important for water absorption by root hair cells. Unlike diffusion, which can take place in gases or in solutions, osmosis only involves solutions.

In addition, while diffusion can happen across cell membranes but does not have to, osmosis only happens across partially permeable membranes such as cell membranes.

Active transport is important for the absorption of mineral ions by plant root hair cells. During active transport, substances move from a more dilute solution to a more concentrated solution. This is against a concentration gradient, whereas diffusion happens down a concentration gradient.

Diffusion and osmosis are passive processes that do not need energy from respiration. However, active transport does need energy from respiration and cannot happen without it.

Command word: Compare

When you are asked to **compare** something, you must describe the similarities and/or differences between two or more things. It is not enough just to write about one of the things.

This part of the answer defines diffusion and gives suitable examples, but does not yet compare it with the other two processes. Carbon dioxide and oxygen could be named as the gases involved.

The answer should have stated that osmosis involves the net movement of water from a dilute to a more concentrated solution.

The use of the connective 'in addition' signals that a related idea is about to follow.

The use of the connective 'whereas' signals that an opposite idea is about to follow.

The use of the connective 'however' also signals that an opposite idea follows.

You need to show comprehensive knowledge and understanding, using relevant scientific ideas, to support your answer. The three transport processes are defined, and suitable examples of each one are given. Their similarities and differences are identified in a logical way.

Now try this

Cells are the basic unit of all animals and plants. Compare the structure of a typical animal cell with the structure of a typical plant cell. In your answer, you should name the main sub-cellular structures and describe their functions. **(6 marks)**

You must provide a written answer, but a labelled diagram can help to make some points clearer.

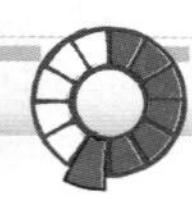

The digestive system

An **enzyme** is a biological catalyst: a protein that can speed up or control a specific reaction.

Digestive enzymes

Digestive enzymes convert larger food molecules into smaller, soluble molecules that can be absorbed into the bloodstream:

1. **Carbohydrase** enzymes
   carbohydrates → simple sugars
   Amylase is produced by the salivary glands, pancreas and small intestine. It breaks down **starch** in the mouth and small intestine.
2. **Protease** enzymes
   proteins → amino acids
   Produced by the stomach, pancreas and small intestine. Breakdown of **proteins** occurs in the stomach and small intestine.
3. **Lipase** enzymes
   lipids → fatty acids + glycerol
   Produced by the pancreas and small intestine. Breakdown of **lipids** (fats and oils) occurs in the small intestine.

The digestive system

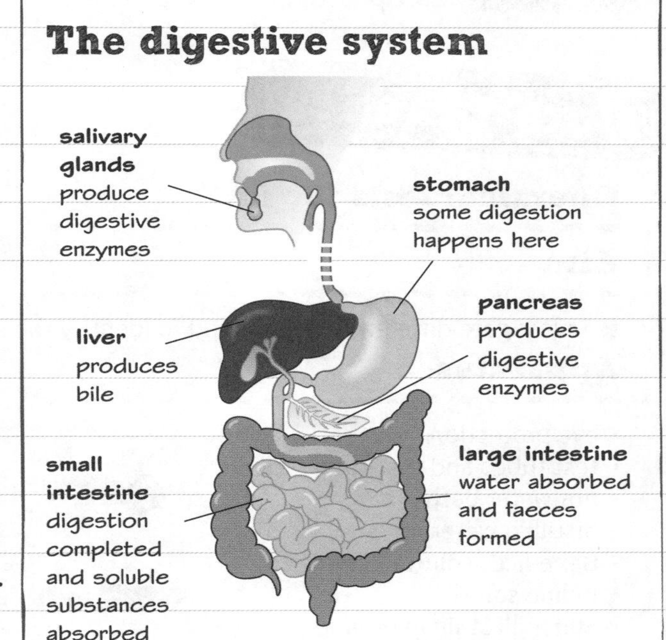

The **digestive system** is an organ system in which several organs work together to digest and absorb food.

Worked example

Describe **two** uses of the products of digestion by the body. **(2 marks)**

Simple sugars can be used to build new carbohydrates.
Some of the glucose produced by digestion of carbohydrates is used in respiration.

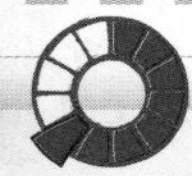

Other products of digestion can be used to build other new substances:
- amino acids for new proteins
- fatty acids and glycerol for new lipids.

You can revise the role of enzymes in metabolism (all the reactions in a cell or body) on page 48.

Principles of organisation

The bodies of large multicellular organisms are organised into different **organ systems.**

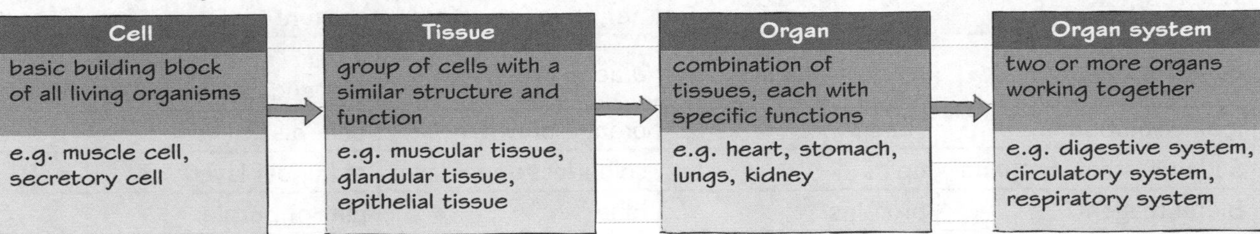

Make a table to summarise the features of amylase, proteases and lipases. In your table, show their sites of production and action, the substances they break down and the products formed. **(6 marks)**

Food testing

You can use **qualitative reagents** to test for a range of carbohydrates, lipids and proteins.

Core practical

Chemical tests for food

Aim

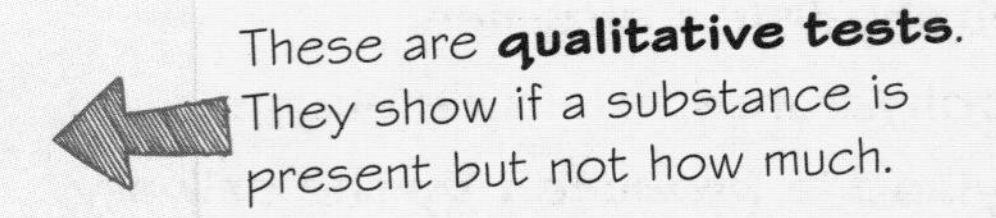

to investigate different types of food to identify the substances they contain

Apparatus

- eye protection
- test tubes and test-tube rack
- hot water bath
- distilled water
- Benedict's solution
- iodine solution
- Sudan III stain in ethanol solution
- Biuret reagent
- food samples

> Make a hot water bath by half filling a beaker with hot (not boiling) water from a kettle.

> Wash spills off clothing or skin immediately.

> Ethanol is highly flammable.

Methods

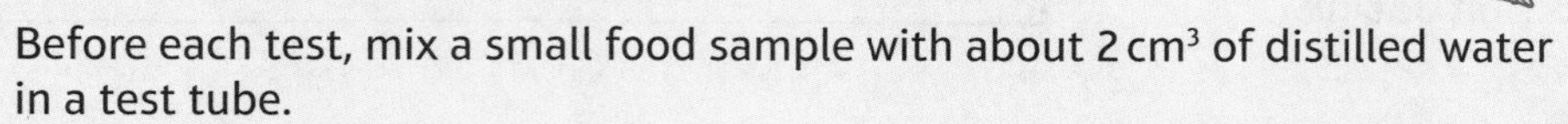

Before each test, mix a small food sample with about 2 cm³ of distilled water in a test tube.

> If you are given larger pieces of food, cut them into small pieces or grind them with a pestle and mortar.

Test for reducing sugars: Add a few drops of Benedict's solution and heat in the hot water bath for 5 minutes. Record any colour change.

> Glucose and fructose are **reducing sugars**, but sucrose (table sugar) is not.

Test for starch: Add a few drops of iodine solution. Record any colour change.

Test for lipids: Add a few drops of Sudan III stain solution and shake to mix. Record whether a red upper layer forms.

> You could just use ethanol instead, and look for a cloudy upper layer.

Test for proteins: Add an equal volume of Biuret reagent and shake to mix. Record any colour change.

Results

Make a table to record your results. This one shows the expected results for each substance.

Test reagent	Substance detected	Colour of reagent or negative result	Colour or result if substance present
Benedict's solution	reducing sugars	blue	green → orange → red (depending on amount of sugar)
Iodine solution	starch	orange-brown	blue-black
Sudan III stain in ethanol	lipids	colourless upper layer	red upper layer
Biuret reagent	proteins	blue	pink or purple

Now try this

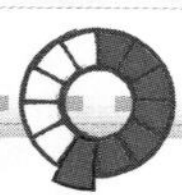

Suggest **two** reasons that explain why a hot water bath is safer to use in this investigation, rather than a Bunsen burner. **(3 marks)**

> Think about how the test for reducing sugars is carried out, and the nature of the substance used in the test for lipids.

Enzymes

The **lock and key theory** is a simple model to explain the action of enzymes.

Lock and key theory

Enzymes are protein molecules, with complex shapes which are important for their activity:

- Part of an enzyme molecule is its **active site**.
- A **substrate** molecule fits into the active site, like a key fits into a lock.
- Other substances do not fit, so are not substrates for the enzyme.
- If the shape of an active site changes, the enzyme is **denatured** and no longer works.

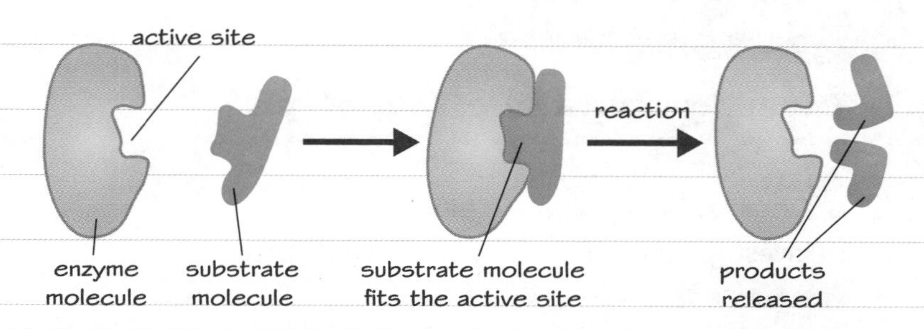

Effect of temperature

The rate of chemical reactions increases as the **temperature** increases. This happens with enzyme-catalysed reactions, but:

- There is an **optimum temperature** at which there is a maximum reaction rate.
- At higher temperatures the active site changes shape, reducing the reaction rate.

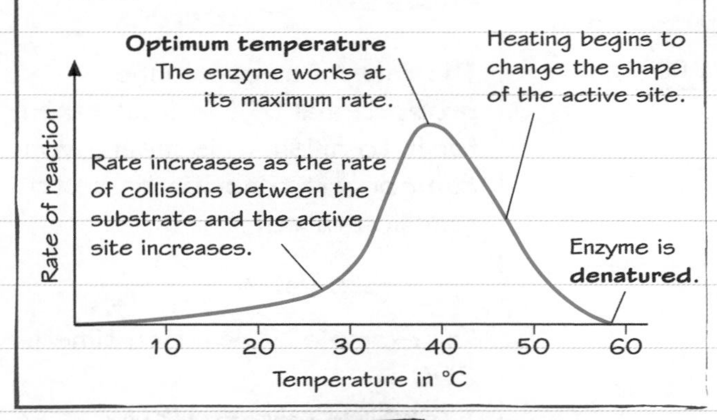

Effect of pH

The **pH** of an enzyme's surroundings affects the shape of the active site. This means:

- Enzymes have an **optimum pH** at which there is a maximum reaction rate.
- Below and above the optimum pH, the bonds in the enzyme are affected, so the shape of the active site changes.

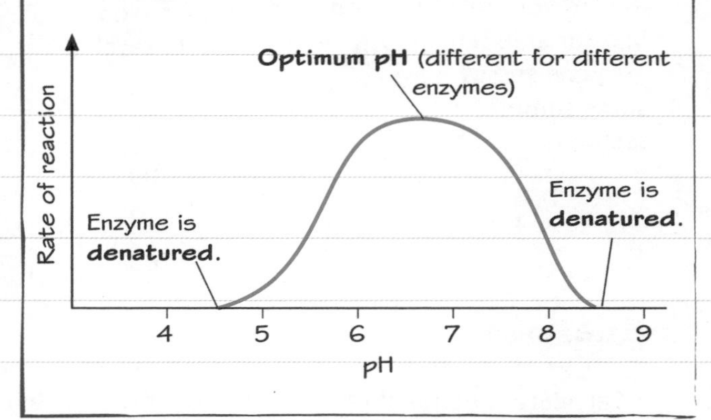

Worked example

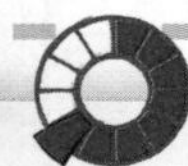

Bile is made in the liver, stored in the gall bladder and released into the small intestine. Explain its role in the digestion of fats. **(3 marks)**

Bile is alkaline, so it neutralises hydrochloric acid made by the stomach. It also emulsifies fat, breaking it into small droplets. This increases the surface area of the fat. The alkaline conditions and the large surface area increase the rate at which lipases break down fat.

Partly digested food is acidic as it reaches the small intestine. The optimum pH for the lipase is pH 7.5–8.0 (alkaline conditions).

Now try this

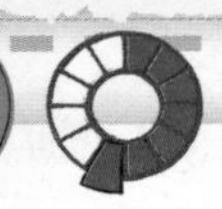

1. Proteases digest proteins in the stomach. Suggest the optimum pH of these proteases, and justify your answer. **(2 marks)**
2. Explain why an enzyme no longer works if the temperature becomes too high. **(2 marks)**

Investigating enzymes

You can use **continuous sampling** to find the time taken to digest starch at different pH values.

Core practical

Investigating the effect of pH

Aim

to investigate the effect of pH on amylase activity

Apparatus and Method

1 Add one drop of iodine solution to each well in a spotting tile.
2 Add the following to a test tube: 2 cm^3 of a pH buffer solution, 2 cm^3 of amylase solution, and 2 cm^3 of starch solution.
3 Mix with a glass rod and start a stop clock.
4 After 30 seconds, use a glass rod to transfer one drop to a well in the spotting tile. Rinse the glass rod with water.
5 Repeat step 4 until there is no change in the colour of the iodine solution. Stop the stop clock and record the time.
6 Repeat steps 1 to 5 using different pH buffer solutions.

Results

Record your results in a suitable table. This one shows some sample results.

pH of buffer	Time taken to digest starch (s)	Rate of reaction (/s)
4	169	0.0059
5	125	
6	100	
7	118	
8	260	

Analysis

1 Calculate the rate of reaction that corresponds to each reaction time.
2 Plot a graph with:
- rate of reaction on the vertical axis
- pH of buffer solution on the horizontal axis.

3 Use your results to identify the **optimum pH** for amylase activity.

Amylase is an enzyme that breaks down starch into simple sugars.

All the solutions should be kept at the same temperature. You can achieve this by keeping the bottles of solution in a water bath maintained at 25 °C.

When the iodine in a well no longer changes colour, there is no starch present – it has all been digested to form simple sugars.

A **buffer solution** is a mixture that resists changes to its pH.

Rate of reaction

The rate of a reaction is proportional to the time taken for it to finish. This means you can convert reaction times to reaction rates:

$$\text{rate (/s)} = \frac{1}{\text{time (s)}}$$

For example, if reaction time is 200 s:
- reaction rate = 1/200 = 0.005 /s

Choose sensible scales that allow the plotted points to occupy at least 50% of the area of the graph. Draw a curve of best fit.

The optimum pH is the pH that gives the greatest rate of reaction.

Now try this

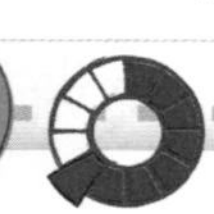

1 Calculate the rates of reaction for the reaction times shown in the table. Give your answers to 2 significant figures. **(2 marks)**
2 Draw a suitable graph to show the rates of reaction between pH 4 and 8. **(3 marks)**
3 Use your graph drawn for question 2, or the table of results, to identify the optimum pH for amylase. Explain your answer. **(2 marks)**
4 Explain why the temperature is controlled in this investigation. **(2 marks)**

The blood

Blood is a tissue with red blood cells, white blood cells and platelets suspended in plasma.

Plasma

Blood **plasma** is a yellow liquid. It transports cells and platelets around the body through the **blood vessels**. It also transports dissolved substances, including:

- carbon dioxide from respiring cells
- soluble products of digestion, such as glucose and amino acids
- urea – a waste product formed when the liver breaks down excess proteins.

Centrifuging blood

The different parts in blood settle out quickly when blood is spun in a centrifuge.

plasma (55%)

white blood cells and platelets (<1%)

red blood cells (45%)

Red blood cells

Most of the cells in blood are **red blood cells**. These cells transport oxygen around the body through the blood vessels. They have several adaptations for this function.

Adaptation	Reason
contain **haemoglobin**	red pigment which binds reversibly to oxygen
no nucleus	provides more space to contain more haemoglobin
biconcave disc shape	increases the surface area to volume ratio for diffusion

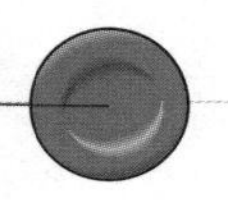

White blood cells

White blood cells are larger than red blood cells and have a **nucleus**. They are part of the body's **immune system**. There are different types of white blood cell, including:

- **phagocytes**, which engulf and destroy **pathogens** (disease-causing organisms)
- **lymphocytes**, which produce antibodies – proteins that bind to pathogens and lead to their destruction.

Platelets

Platelets are fragments of cells with no nucleus. They are involved in forming a blood clot at a cut or wound. Platelets are trapped when a meshwork of fibrin protein forms.

Worked example

The photo shows a thin smear of blood, viewed through a light microscope. Identify A, B and C. **(3 marks)**

A is a platelet, B is a white blood cell and C is a red blood cell.

Platelets are much smaller than red cells or white cells. White cells have a nucleus (the dark area inside) but red cells do not. Blood contains more red cells than white cells.

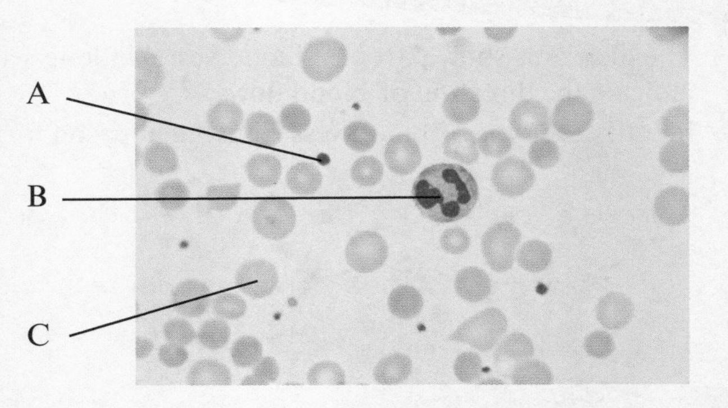

A blood smear stained with a dye.

Now try this

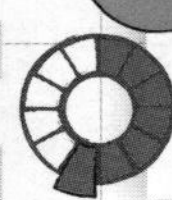

Compare the structure of red blood cells with the structure of white blood cells. **(4 marks)**

Blood vessels

The body contains three different types of **blood vessel**, called arteries, veins and capillaries.

Arteries

Arteries carry blood **A**way from the heart. The blood in arteries is under high pressure:

- This increases each time the heart beats.

Arteries have thick walls containing muscle and elastic fibres. These allow arteries to:

- stretch, without bursting, as blood is pushed through them
- regain their shape afterwards.

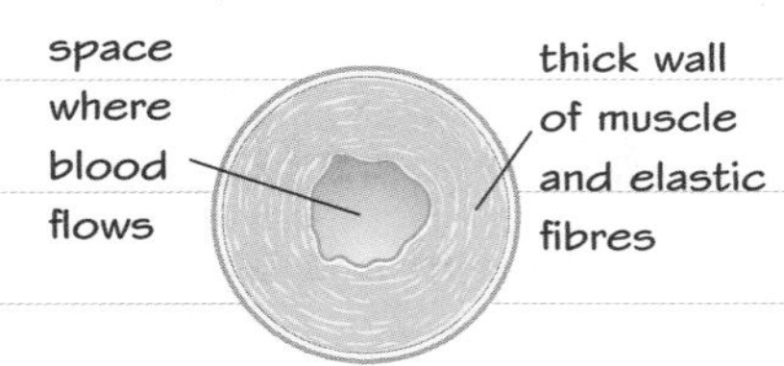

Artery (cross-section)

Veins

Veins carry blood towards the heart. The blood in veins is under the lowest pressure, and is not pumped by the heart. Veins have:

- thinner walls than arteries
- a large **lumen** (space inside)
- **valves** to prevent blood flowing backwards.

Blood is squeezed through the veins by the contractions of surrounding skeletal muscle.

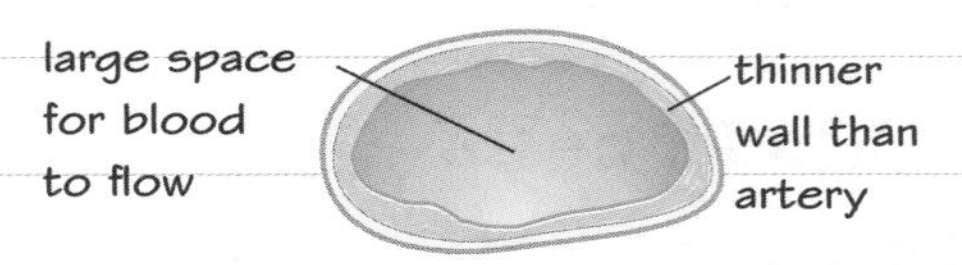

Vein (cross-section)

Capillaries

The **capillaries** are the smallest blood vessels, and contain blood at low pressure.

They form capillary beds – networks that supply blood to the body's tissues and organs.

Substances are exchanged between the blood in the capillaries and the cells around them.

Capillaries:

- are typically one red blood cell wide
- have thin walls, one cell thick, allowing substances to diffuse easily through them.

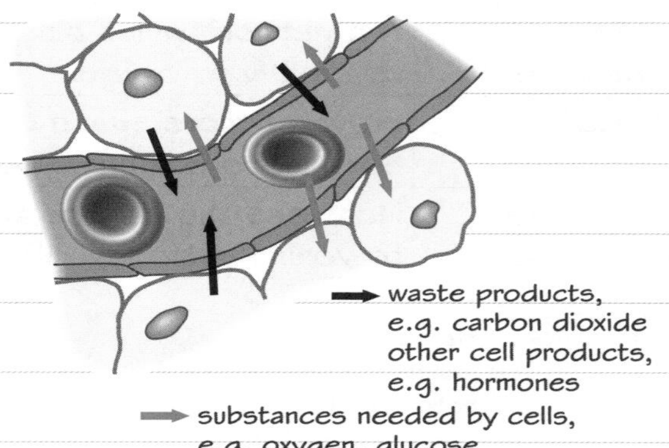

Worked example

The diagrams show part of a blood vessel in long section. The arrows indicate the direction of blood flow.
Identify the type of blood vessel, giving a reason for your answer. **(2 marks)**

It shows a vein because there are valves, which are found in veins.

Blood can only flow one way between the flaps in the valve.

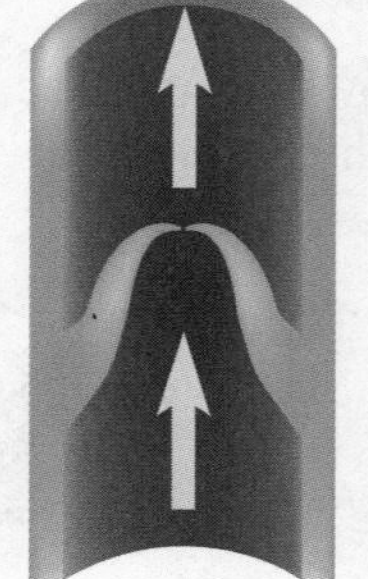

Now try this

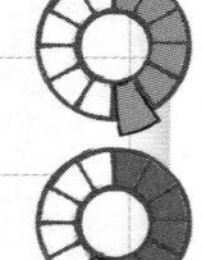

1 List the three types of blood vessel, in order of decreasing blood pressure. **(1 mark)**

2 Explain why almost every body cell is very close to a capillary. **(3 marks)**

3 Give **two** reasons that explain why the walls of arteries are thick, and have muscle and elastic fibres. **(2 marks)**

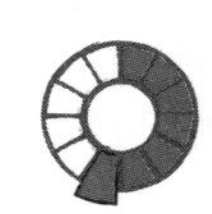

The heart

The **heart** is an organ that pumps blood around the body.

Structure of the heart

Cells in the heart are supplied with oxygenated blood by the **coronary arteries**.

The diagram shows other structures in the heart.

Pulmonary artery carries deoxygenated blood from heart to lungs.

Aorta carries oxygenated blood from heart to body.

Vena cava brings **deoxygenated blood** from body to heart.

Pulmonary vein brings oxygenated blood from lungs to heart.

right atrium

left atrium

Valves prevent blood flowing wrong way through heart (**backflow**).

left ventricle

right ventricle

Left ventricle muscle wall is thicker than right ventricle as it pushes blood all round the body.

■ deoxygenated blood

■ oxygenated blood

The sides of the heart are labelled left and right as if you were looking at the person. So the left side of the heart is on the right side of the diagram.

Remember - arteries take blood away from the heart, veins bring it back in to the heart.

Blood enters the atria. → The atria contract, forcing blood into the ventricles. → The ventricles contract, forcing blood into the arteries. → Blood flows through arteries to the organs and returns to the heart through veins.

Double circulation

Humans have a **double circulatory system**:

- The right ventricle pumps blood to the lungs where gas exchange takes place.
- The left ventricle pumps blood around the rest of the body.

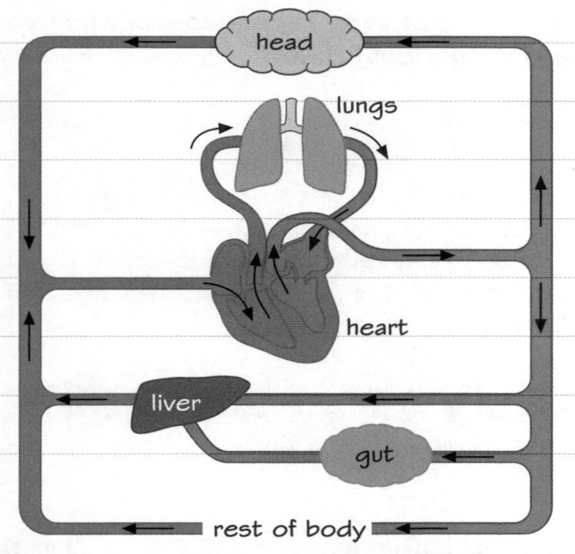

Worked example

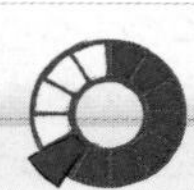

The natural resting heart rate is controlled by a group of cells located in the right atrium that act as a pacemaker.

A man's resting heart rate is 72 beats/min, and his stroke volume is $0.075\ dm^3$/beat.

Calculate the cardiac output in dm^3/min.

(2 marks)

cardiac output = stroke volume × heart rate

$= 0.075 \times 72 = 5.4\ dm^3/min$

Heart rate is the number of heartbeats per minute. The stroke volume is the volume of blood pushed into the aorta in one beat.

Now try this

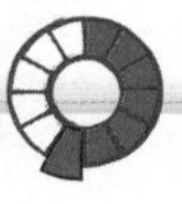

Compare the functions of the left and right ventricles. **(2 marks)**

The lungs

The **lungs** are part of the **respiratory system**, and are adapted for gaseous exchange.

Structure of the lungs

Air enters the mouth or nose and passes through the **trachea** (wind pipe):

- The trachea branches into two **bronchi**, one **bronchus** for each lung.
- These branch further and end in structures called the **alveoli**.

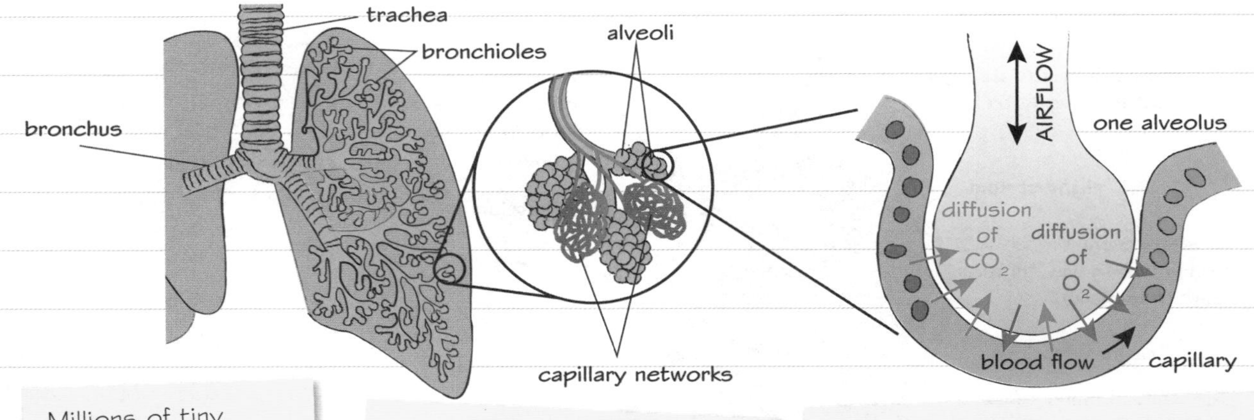

Millions of tiny **alveoli** (air sacs) create a large surface area for **diffusion** of gases.

Each alveolus is closely associated with a **capillary**. Their walls are one cell thick, minimising diffusion distance.

Ventilation of alveoli (by breathing) and continual blood flow through capillaries maintains high concentration gradients, to maximise rate of diffusion.

Gas exchange

Cells need oxygen for **respiration**:

- It is carried in the red blood cells, and diffuses from the capillaries into cells.

Carbon dioxide is a waste product:

- It diffuses from cells into the blood plasma in the capillaries.

In the alveoli in the lungs:

- Oxygen diffuses from the alveolar air into the blood.
- Carbon dioxide diffuses from the blood into the alveolar air.

Inhaled and exhaled air

Inhaled air is different to exhaled air:

Gas	% in inhaled air	% in exhaled air
oxygen	21	16
carbon dioxide	0.04	5

The surface of the alveoli is moist: exhaled air contains more water vapour than inhaled air. You can see this water vapour condensing when you breathe out on a cold day.

Worked example

Explain how the shape of the alveoli is an adaptation to their function. **(3 marks)**

Alveoli have a roughly spherical shape. This allows them to contain a large volume of air within a relatively large surface area.

Multicellular organisms need exchange surfaces with large surface area to volume ratios. The greater the surface area, the greater the rate of diffusion (alveoli give the lungs an area of 50–75 m^2).

Now try this

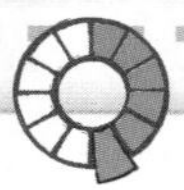

Describe **three** ways in which exhaled air differs in composition from inhaled air. **(3 marks)**

Cardiovascular disease

Cardiovascular disease is a general term for conditions affecting the heart or blood vessels.

Coronary heart disease

Fatty material can build up inside the **coronary arteries**, the arteries that supply blood to the heart itself. This reduces:

- blood flow in the arteries
- the amount of oxygen for the heart muscle
- it also causes pain, heart attacks or death.

A narrowed artery can be treated using a **stent**. This is a wire frame inserted into the damaged artery. It is expanded using a tiny balloon, which is then removed allowing blood to flow more freely.

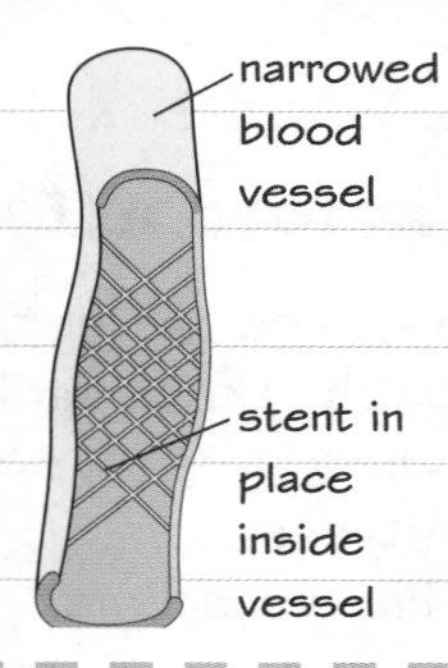

Types of treatment

Cardiovascular diseases are **non-communicable diseases**. They are not caused by an infectious pathogen, so they cannot be passed from person to person.

Cardiovascular diseases usually get slowly worse over a long time. Treatments include:

- drugs
- artificial devices
- transplants.

An artificial pacemaker is a device implanted under the skin. It sends electrical impulses through a wire to the heart. This corrects irregularities in heart rate.

Faulty heart valves

The **heart valves** stop blood flowing in the wrong direction through the heart. If they become faulty, the heart becomes less efficient at pumping blood to the lungs and the rest of the body. Faulty valves may:

- not open fully
- develop a leak.

This causes tiredness and breathlessness, and can lead to death.

Faulty heart valves can be replaced by:

- **mechanical valves** made from metals or polymers
- **biological valves**, from human donors, cows or pigs.

Patients have to take drugs to prevent blood clots caused by mechanical valves and hearts, or to prevent rejection of biological valves by the immune system.

Heart failure

Heart failure is when the heart can no longer pump enough blood for the body's needs. Treatment can include:

- heart **transplant**
- heart and lungs transplant.

Artificial hearts are sometimes used to keep patients alive until their heart transplant, or to rest the heart to help recovery. Treatments like this are expensive, with significant risks.

Worked example

Statins are drugs that reduce blood cholesterol levels. Explain why this decreases the risk of developing heart disease or suffering a heart attack. **(2 marks)**

If the cholesterol level is reduced, the rate at which fatty material builds up in the arteries is also reduced. This means that the coronary arteries are less likely to become blocked.

Statins are relatively cheap drugs. However, they can produce side effects, such as muscle pain and an increased risk of diabetes.

Statins and other drugs have drawbacks as well as benefits. The results of research suggest that the benefits greatly outweigh the risks of harm.

Now try this

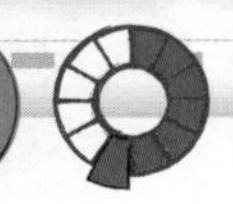

Describe the changes that happen to the coronary arteries that lead to coronary heart disease. **(3 marks)**

Health and disease

Diseases are major causes of ill-health.

What is health?

Health is the state of physical and mental wellbeing.

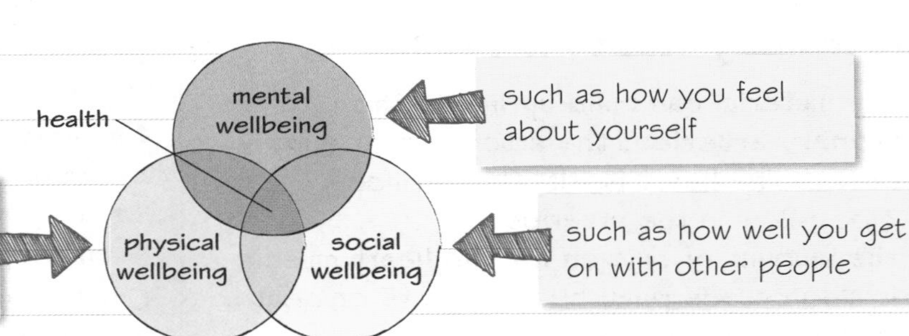

Types of disease

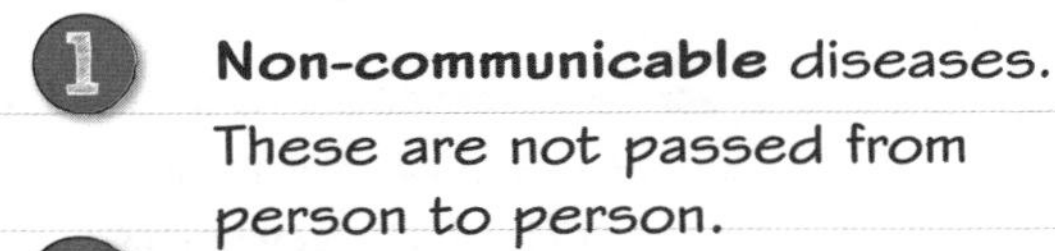

1. **Non-communicable** diseases.
   These are not passed from person to person.
2. **Communicable** diseases.
   These can be passed from one person to another. You can revise pathogens and the diseases they cause on pages 34–36.

Non-communicable	Communicable
number of cases changes gradually over time	rapid variation in the number of cases over time
cases may be widespread	cases are often localised
e.g. cardiovascular disease, cancer, diabetes	e.g. colds, flu, measles, food poisoning

Factors affecting health

Diseases are not the only factors that affect health. These factors also affect health:

- diet
- stress
- life situations.

Diseases may also interact with one other.

Disease or problem	Additional problem
defects in the immune system	more prone to infectious diseases
immune reaction caused by a pathogen	can trigger allergies, e.g. skin rashes, asthma
severe physical ill-health	can lead to mental illness such as depression
viruses in cells	can be a trigger for cancer

Worked example

Helicobacter pylori is a bacterium that lives in the stomach lining. People usually become infected during childhood, probably from family members. About 20% of people in the UK have *H. pylori*. Of these, about 15% develop stomach ulcers. Long-term use of high doses of aspirin also cause stomach ulcers. Evaluate whether stomach ulcers are a communicable disease. **(5 marks)**

H. pylori is caught from other people. If most people with ulcers also have this bacterium, stomach ulcers would be a communicable disease. On the other hand, most infected people do not develop ulcers, and people can also develop ulcers because they take a lot of aspirin over a long time. It appears that stomach ulcers might be a communicable disease, but not enough information is given to be sure.

A course of antibiotics kills *H. pylori* and most ulcers heal in 1–2 months. If ulcers are caused by aspirin, doctors can prescribe other painkillers.

Now try this

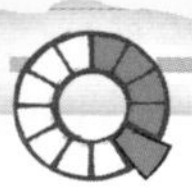

1 Give **one** example of a non-communicable disease. **(1 mark)**
2 Give **four** factors that may cause ill-health. **(4 marks)**

Lifestyle and disease

Lifestyle factors such as diet are linked to increased rates of disease.

Exercise and disease

Obesity (being very overweight) is a lifestyle risk factor for type 2 **diabetes** (you can revise diabetes on page 59). A person who exercises regularly is more likely to control their body mass. They are also less likely to suffer from **cardiovascular disease** (see page 25).

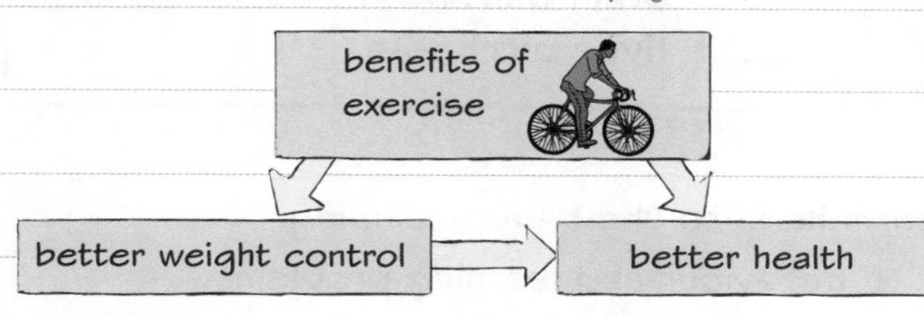

Inherited factors

Inherited factors are not lifestyle factors, but they can affect health. For example, some people inherit certain **alleles** that increase their blood cholesterol level. They may have higher cholesterol levels than people without these alleles and who eat the same diet.

When we study the effects of lifestyle factors on disease, we must compare people with similar inherited factors to obtain valid results.

Some other risk factors

Many diseases are caused by the interaction of different factors. Some risk factors have a proven **causal mechanism**. Scientists:

- can show that the risk factor is a cause of the disease, and
- can show how this works.

For example, risk factors with proven causal mechanisms for cancers include:

- toxic substances called **carcinogens**
- exposure to ionising radiation.

Scientists have also identified lifestyle risk factors and genetic risk factors for cancers.

Cancer

Cancer is caused by changes in cells that lead to uncontrolled cell growth and division.

1. **Benign tumours** are growths of abnormal cells. These cells are contained in one area, usually inside a membrane. They do not invade other parts of the body.
2. **Malignant tumours** are cancers. They invade neighbouring tissues. They also spread to different parts of the body in the blood, forming **secondary tumours**.

Worked example

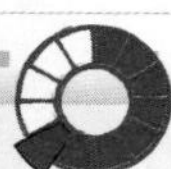

In a study, 522 overweight people of similar age who were at risk of developing diabetes were separated into two groups.

Group A was given advice on how to live more healthily. Group B, the control group, received no such advice. After four years, 11% of Group A and 23% of Group B had developed diabetes.

It was concluded that living healthily reduces the risk of developing diabetes. Explain the extent to which the results support this conclusion.

(3 marks)

The results support the conclusion. Only 11% of the group who received advice developed diabetes, just under half of the percentage in the control group. The results are likely to be valid because the study used many people, of similar age, and who all started with the same problems.

The validity of a study increases as the number of people involved increases, and by having a control group.

Now try this

A study into the effects of physical fitness found that unfit overweight men had a much higher risk of type 2 diabetes than fit overweight men.

Suggest another suitable group of men for this study.

(1 mark)

Alcohol and smoking

Drinking **alcohol** and **smoking** are lifestyle risk factors for some non-communicable diseases.

Effects of alcohol

Ethanol is the alcohol found in alcoholic drinks.

Drinking a lot of alcohol at one time can cause **short-term harm** including:
- blurred vision
- slow reactions
- lowered inhibitions so you take more risks.

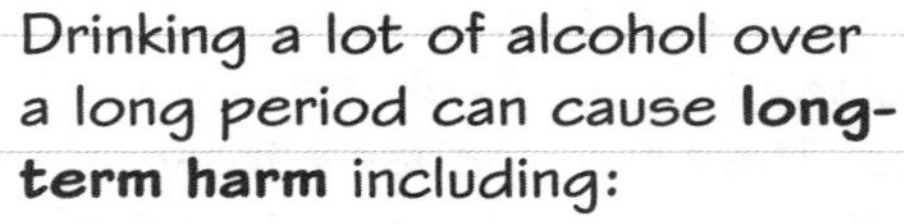

Drinking a lot of alcohol over a long period can cause **long-term harm** including:
- brain damage
- **liver cirrhosis**.

Drinking alcohol while pregnant can lead to long-term harm to the baby, including:
- poor growth
- abnormal facial features
- behaviour and learning problems.

The risk increases as consumption increases.

The effects of smoking

Tobacco smoke contains many different harmful substances.

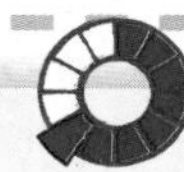

This can lead to low birth weight in babies whose mothers smoke.

Carbon monoxide reduces how much oxygen the blood can carry.

Nicotine is addictive.

Substances in cigarettes can cause blood vessels to narrow, increasing blood pressure. This can lead to **cardiovascular diseases** such as heart attack or strokes.

Chemicals in **tar** are **carcinogens** that cause cancers, particularly of mouth and lungs.

Worked example

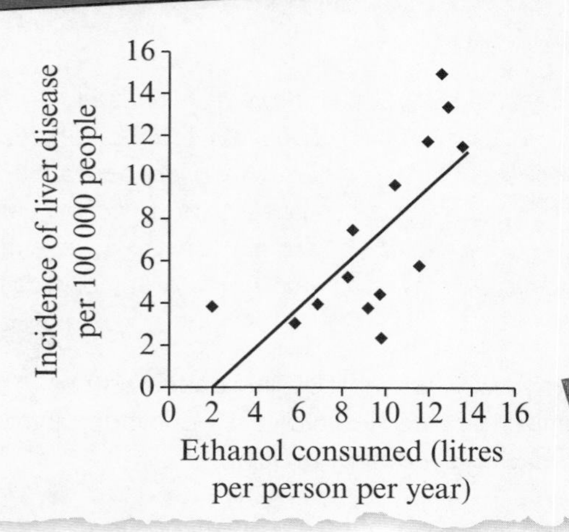

The graph shows the incidence of liver disease and the mean ethanol consumption per person per year for different countries. Describe the relationship and give a reason that explains it. **(2 marks)**.

As the amount of alcohol consumed increases, the incidence of liver disease also increases. This is because ethanol is poisonous, particularly to liver cells.

This graph shows a **positive correlation** between two factors. You might also see a graph where one factor increases while the other decreases – a **negative correlation**.

Now try this

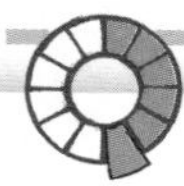

1 Give **two** effects of long-term excessive alcohol consumption. **(2 marks)**

2 Explain why smoking increases the risk of having a heart attack. **(2 marks)**

The leaf

The leaf is a plant **organ** that contains different tissues.

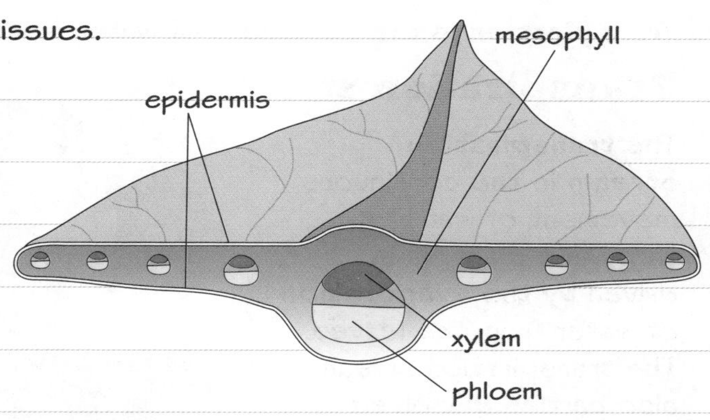

Plant organs

Plant organs include:

- the **leaf**, which carries out photosynthesis
- the **stem**, which supports leaves and flowers
- the **root**, which anchors the plant in the soil, and absorbs water and mineral ions from the soil water.

These organs form a plant **organ system** for transport of substances around the plant.

Leaf tissues

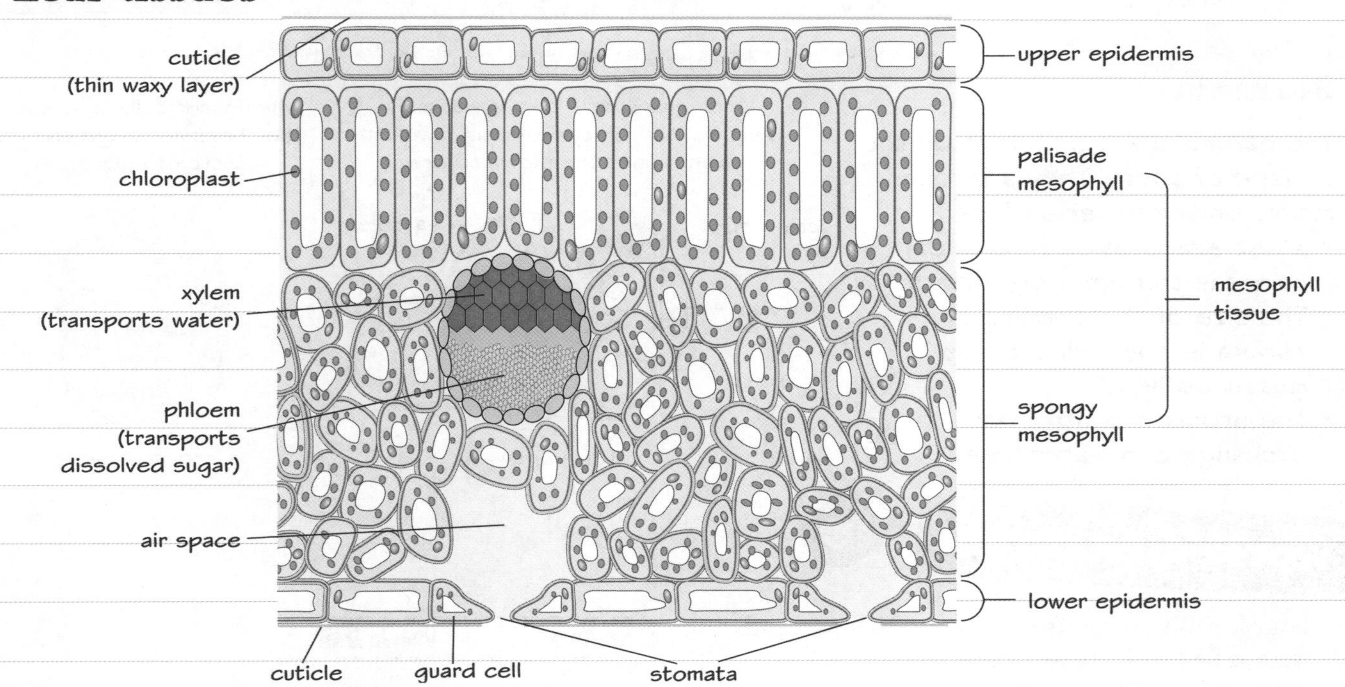

Worked example

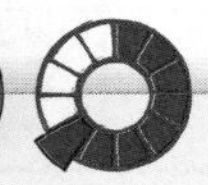

Describe how the mesophyll tissue is adapted for photosynthesis. **(4 marks)**

The palisade mesophyll cells contain many chloroplasts. Their cylindrical shape means that light has few cell walls to pass through to reach the chloroplasts.

The spongy mesophyll cells pack together in an irregular way so that there are air spaces, with large surface areas for gas exchange.

The mesophyll is sandwiched between the upper and lower epidermis. These tissues consist of thin, flat cells that protect the leaf but easily let light through.

Spongy mesophyll cells have fewer chloroplasts than palisade mesophylls. They are found on the lower surface of the leaf, which receives less sunlight.

Now try this

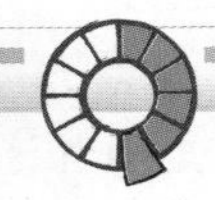

Name **four** tissues found in the leaf. **(4 marks)**

Transpiration

Transpiration is the movement of water from roots to leaves, where it evaporates into the air.

Transpiration stream

The **transpiration stream** is the continuous movement of water through a plant. It is driven by the **evaporation** of water from the leaves. The transpiration stream also carries dissolved **mineral ions.** You can revise adaptations of root hair cells and xylem tissue on page 5.

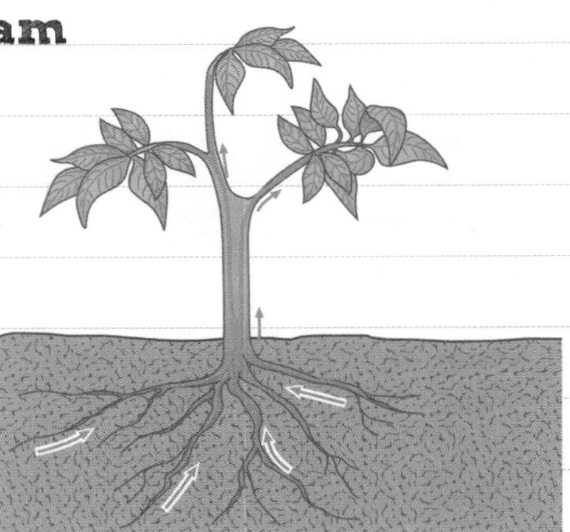

Water is lost by evaporation from the leaves, mainly through the stomata.

↑

Water travels through the stem and into the leaves.

↑

Water enters the xylem and moves from the roots to the stem.

↑

Water enters root hair cells by osmosis. Mineral ions enter by active transport.

Stomata

The **stomata** are openings on the surface of a leaf. They are found mainly on the lower surface:

- Most evaporation from a leaf happens through its stomata.
- The size of the opening in a **stoma** is controlled by its **guard cells.**
- The stomata control gas exchange and water loss.

When guard cells take in water by osmosis, they become turgid (swollen) which causes the stoma to open.

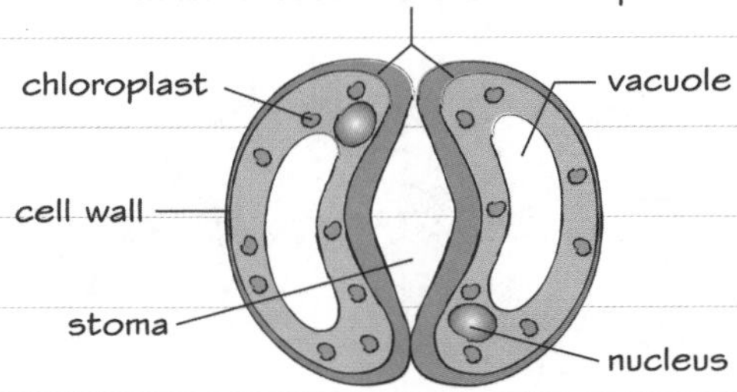

stoma open

When guard cells lose water, they become flaccid and the stoma closes.

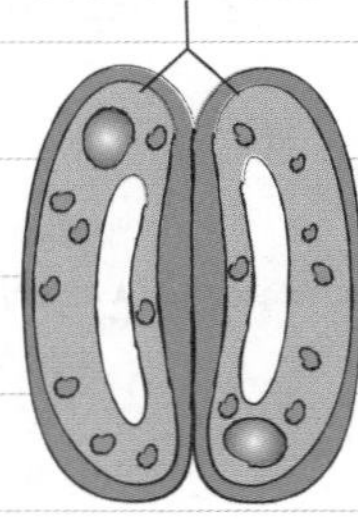

stoma closed

Worked example

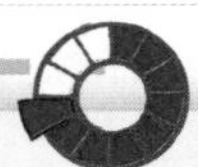

Four leaves were cut from a plant. Their cut ends were sealed with grease. The leaves were covered in grease in the following ways.

- A – Neither surface was greased.
- B – The upper surface was greased.
- C – The lower surface was greased.
- D – Both surfaces were greased.

The mass of each leaf was measured. They were left for the same time before measuring their mass again. The table shows the results.

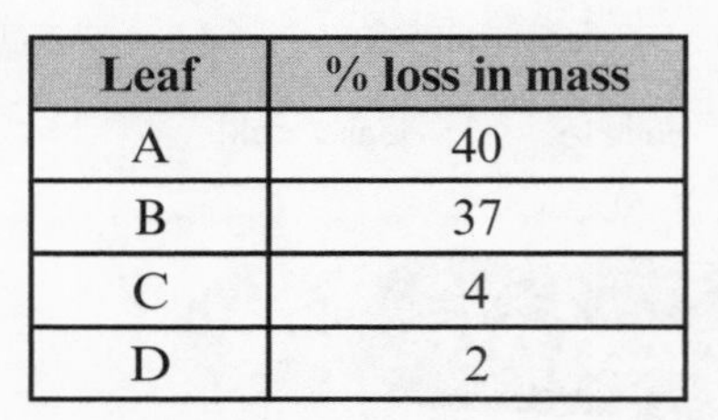

Leaf	% loss in mass
A	40
B	37
C	4
D	2

Explain these results. **(4 marks)**

The loss in mass is due to evaporation of water. Leaves A and B lost the most water vapour because it could still escape through the stomata on the lower surface. Leaves C and D lost very little water because their stomata were covered.

Environmental factors also affect the rate of transpiration (see page 31).

Now try this

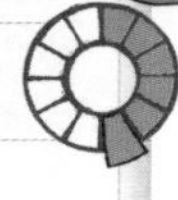

1 Describe how water moves into a plant and escapes from the leaf. **(4 marks)**

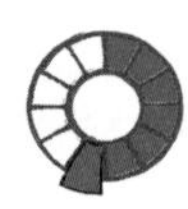

2 Stomata become more open in the daytime and close when it is dark at night. Describe the effect of this change on the rate of transpiration. **(2 marks)**

Investigating transpiration

Environmental factors affect the rate of transpiration. You can measure the **transpiration rate** using a **potometer**.

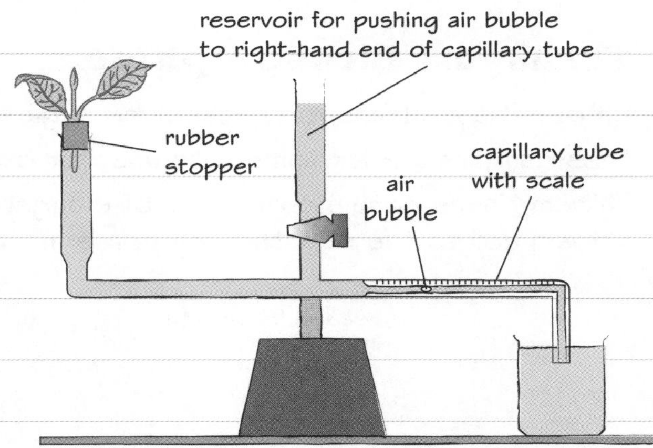

1. Record the starting position of the bubble in the capillary tube using the scale.

2. Record the position of the bubble after allowing the plant shoot to transpire for a known time.

3. Calculate the rate at which the bubble moved. This gives the rate of water uptake, which gives the rate of water loss and so the rate of transpiration.

Factors affecting transpiration rate

Factor	Effect on transpiration rate if factor increases	Explanation of effect
light intensity	increases	• Rate of photosynthesis increases, so the stomata are more open for gas exchange.
temperature	increases	• Rate of photosynthesis increases, so the stomata are more open for gas exchange. • Water molecules move quickly, so rate of diffusion increases.
air movement	increases	• Concentration gradient of water vapour increases, so rate of diffusion increases.

Worked example

A plant shoot was placed in a potometer. The position of the bubble was recorded before, and after, leaving the shoot in still air for 5 minutes. The potometer was reset, and the experiment repeated with the shoot enclosed in a plastic bag. The table shows the results.

	Outside bag	Inside bag
Start reading (mm)	2	5
End reading (mm)	52	25
Distance (mm)	50	20
Rate (mm/min)	50/5 = 10	20/5 = 4

(a) Complete the table. **(2 marks)**

(b) Explain the results. **(2 marks)**

The humidity inside the bag increases as water vapour escapes from the leaves. As the humidity increases, the concentration gradient of water vapour decreases. This reduces the rate of diffusion, so the transpiration rate decreases.

Maths skills You can calculate the **volume** of water taken up by the shoot, in mm^3, if you know the diameter of the capillary tubing in mm:

volume = distance × πr^2

where radius, $r = \frac{\text{diameter}}{2}$

Now try this

1 State **four** environmental factors that affect the rate of transpiration. **(4 marks)**

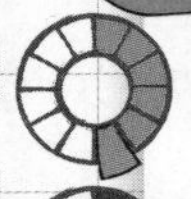

2 A plant wilts and becomes floppy if it loses water faster than the roots can absorb water. Explain why a plant is more likely to wilt when the weather is hot and windy, rather than cool and still. **(5 marks)**

Translocation

Translocation is the movement of dissolved food molecules through phloem tissue.

Transport through phloem

Phloem tissue transports dissolved sugars from the leaves to the rest of the plant.

These sugars are for immediate use, for example in respiration and for storage.

Phloem tissue consists of tubes of elongated cells (see page 5). Cell sap can move from one phloem cell to the next through pores in the end walls.

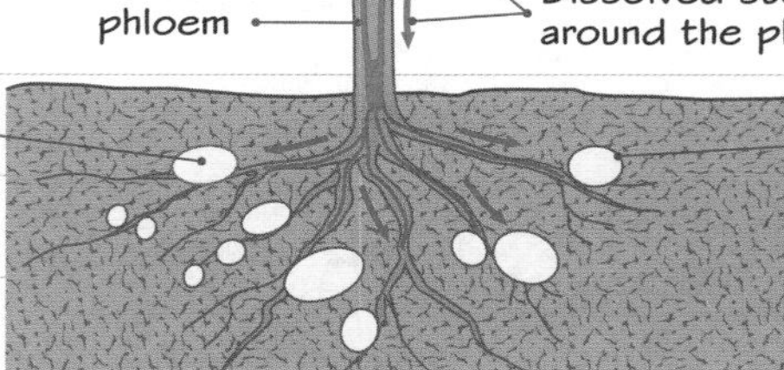

Worked example

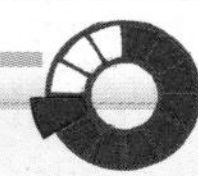

Carbon dioxide containing radioactive carbon atoms is supplied to a plant's leaf. The plant is left in the light for several hours. Its stem is then analysed to detect radioactivity.

Explain why radioactivity is detected in the stem.

(2 marks)

Carbon dioxide is used by the leaf for photosynthesis. This means that the glucose formed contains radioactive carbon atoms. Some glucose is converted to sucrose. This is transported through the plant in the phloem, so radioactivity is detected in the stem.

Some glucose is used in respiration. Some is converted into storage substances such as starch. Radioactive starch would be detected in this plant.

Meristem tissue

Meristem tissue is found at the growing tips of shoots and roots. It consists of **stem cells**.

These cells can **differentiate** to form specialised cells, including those found in xylem and phloem tissue.

You can revise stem cells on page 10.

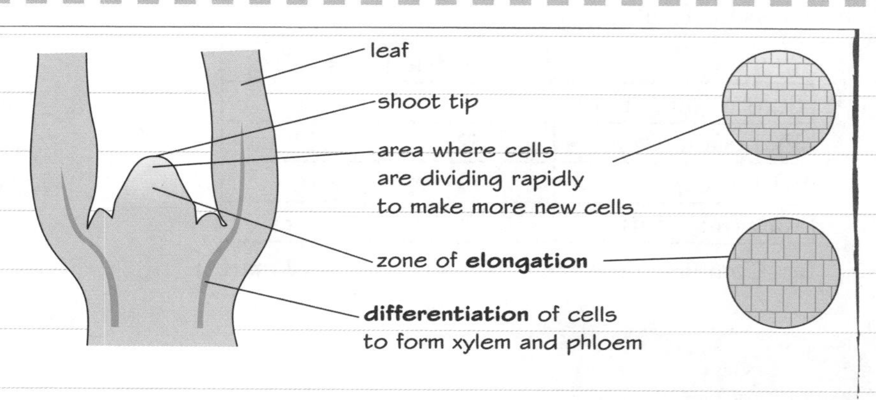

Now try this

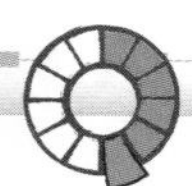

State how sugars move from one phloem cell to the next.

(2 marks)

Extended response – Organisation

There will be at least one 6-mark question on your exam paper. For these questions, you will need to think scientifically and structure your answer logically, showing how the points you make are related to each other.

For the questions on this page, you can revise the circulatory system and the lungs on pages 21–24.

Worked example

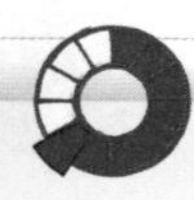

The circulatory system transports substances to and from the tissues in the body.
Explain how this organ system is adapted to supply oxygen and to remove carbon dioxide.

The blood contains red blood cells suspended in plasma. Haemoglobin in the red blood cells binds to oxygen so it can be transported around the body. Carbon dioxide dissolves in the plasma so it can be transported around the body.

The heart has muscles which can contract in an organised way to pump blood. Deoxygenated blood enters the right side of the heart through the vena cava. It is pumped through the pulmonary artery to the lungs. In the lungs, oxygen diffuses from the air into the blood and carbon dioxide diffuses from the blood into the air.

Oxygenated blood returns to the left side of the heart through the pulmonary vein. The heart contains valves so the blood flows in one direction only. The heart pumps blood through the aorta to the rest of the body.

The various arteries carry oxygenated blood to the capillaries in body tissues. The capillaries have thin walls and a large surface area to allow efficient gas exchange between the blood and cells in the tissues.

Command word: Explain

When you are asked to **explain** something, you must include reasoning or justification of the points you make.

Blood also contains white blood cells and platelets, but these have other functions and need not be mentioned.

The answer could also mention that a group of cells located in the right atrium act as a pacemaker. They control the natural resting heart rate.

The answer could also state that this organ system is a double circulatory system. This means it can maintain a high blood pressure and flow rate to the rest of the body.

The answer could also explain that blood leaves the capillaries and enters the veins, so it can return to the heart.

You need to show comprehensive knowledge and understanding, using relevant scientific ideas, to support your answer. Make sure your answer links different points in a logical way.

Now try this

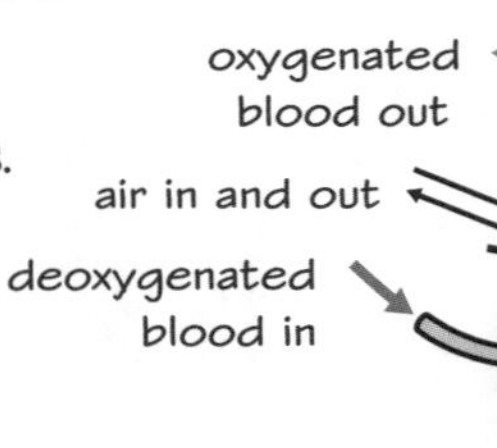

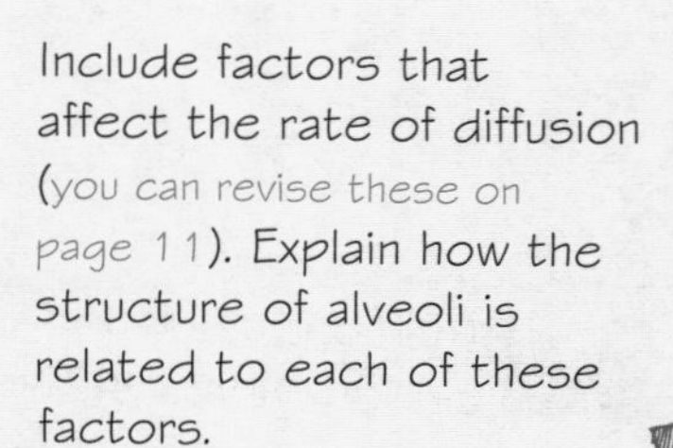

Include factors that affect the rate of diffusion (you can revise these on page 11). Explain how the structure of alveoli is related to each of these factors.

The diagram shows alveoli, structures found in the lungs. Explain how the alveoli are adapted for efficient gas exchange between air in the lungs and blood in the capillaries. **(6 marks)**

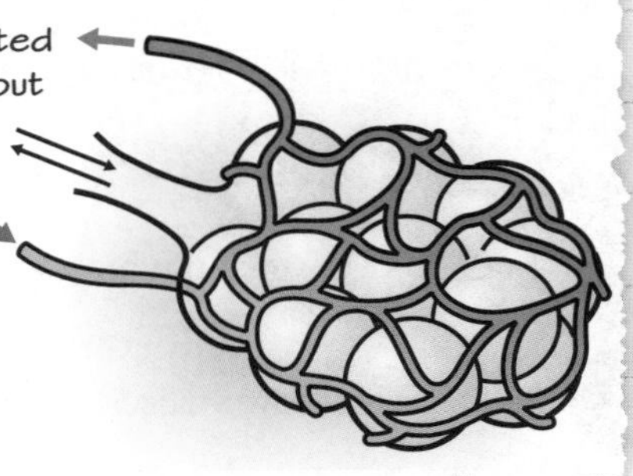

Viral diseases

Pathogens are microorganisms that cause infectious disease. **Viruses** are a type of pathogen.

Viruses

A virus consists of genetic material (DNA or RNA) surrounded by a coat of proteins and lipids.

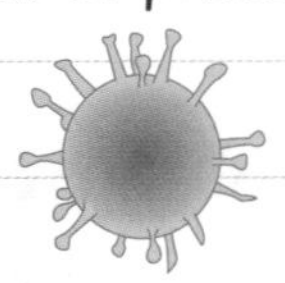

Viruses:
- are much smaller than cells
- exist and reproduce inside cells, causing cell damage
- reproduce rapidly in the body
- can infect plants or animals
- are spread by direct contact, water or air.

Colds and flu are viral diseases.

Virus life cycle

Virus enters the cell through the cell membrane.
↓
Cell produces more viral DNA or RNA, and viral proteins.
↓
New viruses are assembled inside the cell.
↓
New viruses leave or burst out of the cell.

Human viral diseases

The table shows some information about two human viral diseases.

	Measles virus	HIV
How it is spread	• inhalation of droplets from sneezes and coughs	• sexual contact • exchange of body fluids such as blood, e.g. when drug users share needles
Symptoms of infection	• fever and a red skin rash • can be fatal if complications occur (so most young children are vaccinated against measles)	• flu-like illness at the start • attacks body's immune cells unless it is controlled with antiretroviral drugs • AIDS (late-stage HIV infection) occurs when the body's immune system is too badly damaged to deal with other infections or cancers.

Worked example

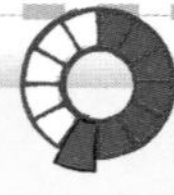

Tobacco mosaic virus, TMV, is a plant pathogen. Describe the symptoms of infection by TMV. **(2 marks)**

The leaves get a distinctive mottled or 'mosaic' pattern of discoloration. This reduces the rate of photosynthesis, so the growth of the plant is affected.

TMV affects tomato plants and many other species of plants, not just tobacco plants.

The virus can spread by contact from cell to cell, and plant to plant. It can also travel through the phloem.

Now try this

1 Give **two** ways in which viruses can be spread. **(2 marks)**

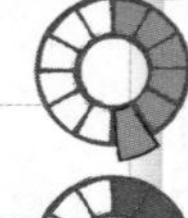

2 Describe how viruses damage their host cells. **(2 marks)**

The host cell is the cell the virus infects.

Bacterial diseases

Bacteria are **prokaryotic** organisms that can act as pathogens.

Bacteria

A bacterial cell contains cytoplasm, a loop of DNA and one or more **plasmids** (rings of DNA) with a cell membrane surrounded by a cell wall.

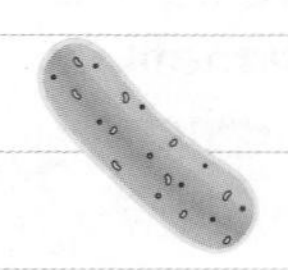

Bacteria:
- are usually smaller than plant and animal cells, but larger than viruses
- may reproduce rapidly inside the body
- may produce **toxins** – poisons that damage tissues, making us feel ill
- can infect plants or animals
- are spread by direct contact, water or air.

Cholera and tuberculosis (TB) are bacterial diseases.

Ignaz Semmelweis (1818–1865)

Semmelweis wondered why many women died of infection soon after childbirth.

Semmelweis realised that doctors might be transferring infection between patients on their hands.

↓

He insisted that doctors wash their hands before examining each patient.

↓

Death rates fell rapidly in the wards where doctors washed their hands.

Pathogens were unknown then, so he could not explain why hand washing worked.

Human bacterial diseases

The table shows some information about two human bacterial diseases.

	Salmonella food poisoning	Gonorrhoea
How it is spread	• food containing bacteria • bacteria on food prepared in unhygienic conditions	• sexual contact
Symptoms of infection	• fever • abdominal cramps • vomiting • diarrhoea	• thick yellow or green discharge from the vagina or penis • pain on urinating

Worked example

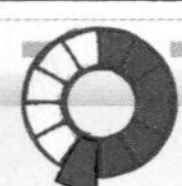

Gonorrhoea is a sexually transmitted disease (STD).

(a) Explain why its spread can be controlled using antibiotics. **(2 marks)**

Gonorrhoea is a bacterial disease and antibiotics can kill infective bacteria inside the body.

(b) Give **one** other way in which the spread of gonorrhoea may be controlled. **(1 mark)**

People could use condoms during sexual intercourse.

Gonorrhoea was easily treated in the past using penicillin, an antibiotic. However, many resistant strains of bacteria have appeared. You can revise antibiotic resistance on page 82.

Around 1 in 10 infected men and almost half of infected women do not have any symptoms. So people may spread the pathogen without knowing they are infected.

Now try this

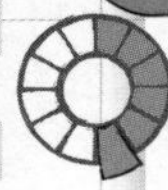

1 Give **one** way in which bacterial toxins cause harm. **(1 mark)**

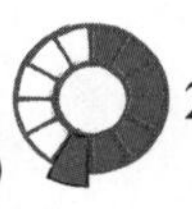

2 Explain why you should wash your hands before handling food. **(2 marks)**

Fungal and protist diseases

Fungi and protists are **eukaryotic** organisms that may act as pathogens.

Fungi

Single-celled fungi include yeasts and moulds. They have cell walls but no chloroplasts. Some may produce toxins that make people ill.

You can revise the use of yeast cells in fermentation to make bread and alcoholic drinks on page 46.

Protists

Protists do not form tissues. They exist as a huge variety of unicellular organisms.

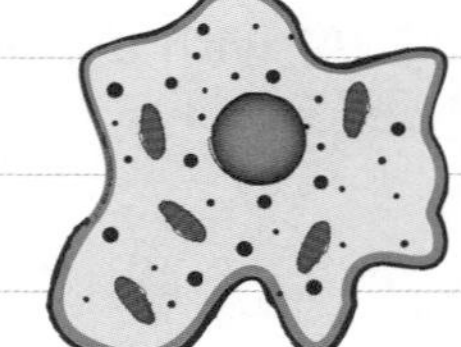

Some protist species are **parasites**: they benefit at the expense of the host they infect.

Malaria

The symptoms of **malaria** include recurring episodes of fever, and death.

The pathogens that cause malaria are protists.

They are parasites with a **life cycle** that involves an insect called a **mosquito**.

The mosquito is a **vector** – it carries the pathogen from person to person.

The spread of malaria is controlled by:

- preventing mosquitoes breeding
- using mosquito nets at night to avoid being bitten.

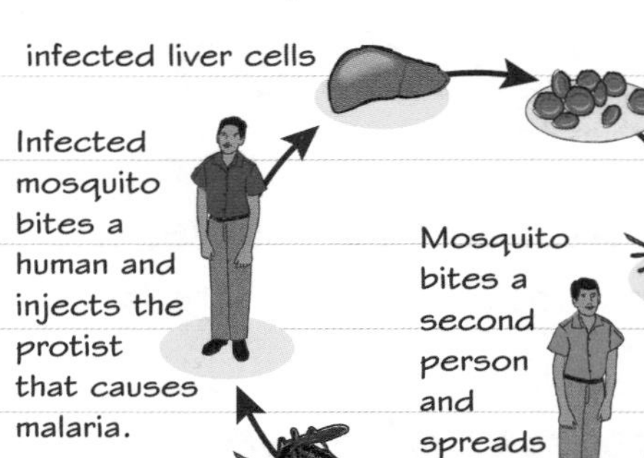

Worked example

Rose black spot is a fungal disease. It causes purple or black spots to develop on leaves. These often turn yellow and drop early.

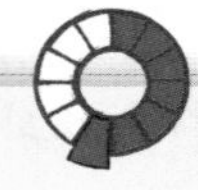

Rose black spot is spread in the environment by fungal spores that travel in the wind, or by water in rain splashes.

(a) Explain why rose black spot affects the growth of infected plants. **(2 marks)**

Photosynthesis is reduced because of the damage to the leaves, so the plant cannot make enough food for healthy growth.

(b) Give **one** way in which rose black spot may be treated. **(1 mark)**

Spray a fungicide onto the plants.

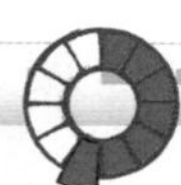

Fungicides are substances that kill fungi. They are used in some treatments for fungal diseases that affect people, such as athlete's foot.

Another way to treat rose black spot is to remove and destroy the affected leaves.

Now try this

1 Ringworm is a fungal disease spread through contact with infected skin, or with contaminated clothes. Suggest **two** ways to prevent its spread. **(2 marks)**

2 Mosquitoes breed in still water. They spread malaria and are controlled in several ways. Give a reason that explains why, in areas affected by malaria:
(a) authorities may drain ditches **(1 mark)**
(b) mosquito nets are placed over beds. **(1 mark)**

Human defence systems

The human body has non-specific defence systems against **pathogens**. These systems are found in various places including the nose, trachea and bronchi, stomach and skin.

Chemical and physical defences

Chemical defences

Lysozyme enzyme in tears kills bacteria by digesting their cell walls.

Lysozyme enzyme is also present in saliva and mucus.

Hydrochloric acid in the stomach kills pathogens in food and drink.

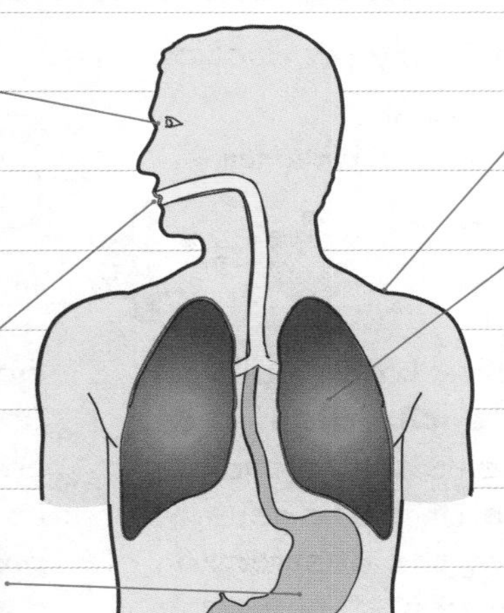

Physical barriers

Unbroken **skin** forms a protective barrier because it is too thick for most pathogens to get through.

Sticky **mucus** in the trachea and bronchi traps pathogens. **Cilia** on the cells lining these passages move in a wave-like motion, moving mucus and trapped pathogens out of the lungs towards the back of the throat where it is swallowed.

Mucus traps pathogens.

Cilia move mucus away from lungs.

cilia

epithelial cells

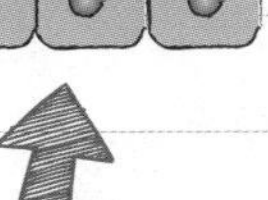

Remember that epithelial cells line the surface of tubes.

Chemical defences involve substances that kill pathogens and make them inactive.

Physical defences make it more difficult for pathogens to enter the body.

Worked example

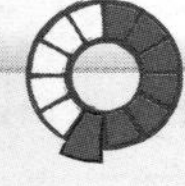

Substances in cigarette smoke paralyse the cilia in the epithelium of the respiratory system. Explain why smokers are more likely to suffer from lung infections than non-smokers. **(3 marks)**

Mucus in the trachea and bronchi traps bacteria. The cilia move this mucus up into the throat, where it is swallowed and destroyed. When the cilia are paralysed, this no longer happens, so bacteria and other pathogens are no longer removed.

Smokers develop a 'smoker's cough'. Their lungs become clogged with mucus which they have to cough up.

Skin defences

The skin is the organ that covers the outer surface of the body. It acts as a barrier to pathogens and has other features that defend against them:

1. It secretes antimicrobial substances that kill pathogenic bacteria or inhibit their growth.
2. Many species of non-pathogenic bacteria live on the skin. They secrete substances that kill pathogenic bacteria, and compete with them for nutrients.
3. Scabs form over damaged skin, keeping pathogens out while the skin repairs itself.

Now try this

1 Give **two** examples of physical defences against pathogens. **(2 marks)**

2 Give a reason that explains why food poisoning might be more likely when a person swallows their food without chewing it thoroughly first. **(1 mark)**

What defences are present in the mouth and stomach?

The immune system

If a pathogen enters the body, the **immune system** tries to destroy the pathogen.

White blood cells

White blood cells help to defend against **pathogens** in different ways.

phagocytosis

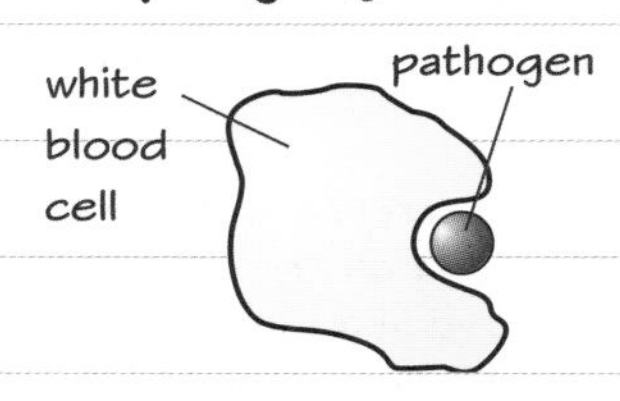

Some white blood cells **ingest** pathogens. They flow around them and destroy them.

antibody production

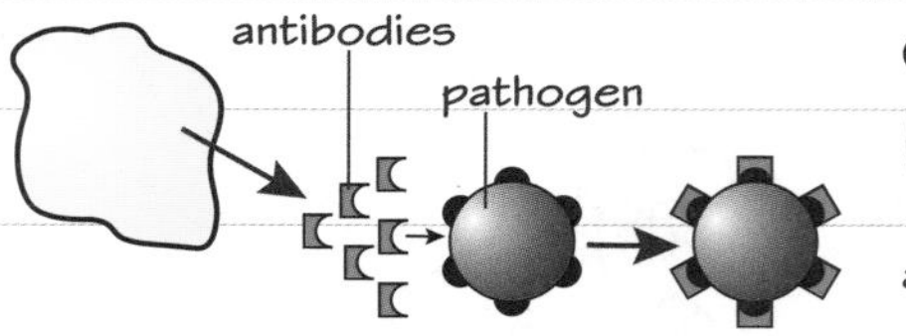

Some white blood cells produce **antibodies**. These are proteins that attach to **antigens** on pathogens, leading to the destruction of the pathogen.

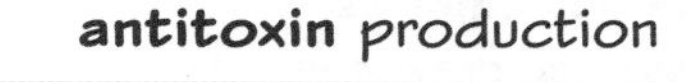

antitoxin production

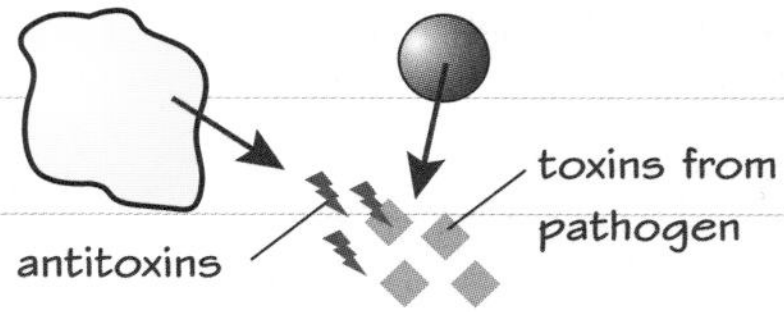

Some white blood cells produce **antitoxins**. These are proteins that attach to poisonous substances produced by pathogens, making them inactive.

Antibodies

Antibodies recognise and attach to antigens (substances produced by organisms):

1. A white blood cell with antibodies specific to an antigen attaches to that antigen.
2. The white blood cell is activated, and divides many times to produce **clones**. These cells produce lots of antibodies against the antigen.
3. Some of the white blood cells remain in the blood, as 'memory lymphocytes', when the infection is over.

Immunity

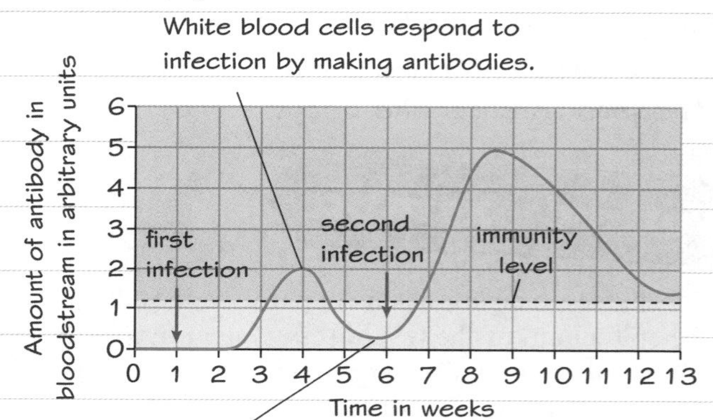

The memory lymphocytes are made after the infection and then respond more quickly to another infection by the pathogen.

An antibody is **specific** for a particular antigen.

Worked example

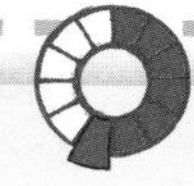

Measles is a viral disease. Explain why you do not get measles more than once. **(2 marks)**

After the first infection with the measles virus, memory cells remain in the blood. If the virus infects the body again, these white blood cells cause a faster and larger immune response so you do not fall ill.

The immune response to the first infection is called the **primary response**. The immune response to a later infection is called the **secondary response**.

Now try this

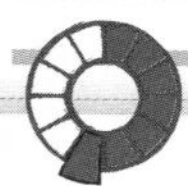

Describe **three** ways in which white blood cells defend the body against pathogens. **(6 marks)**

Vaccination

Vaccination prevents illness in individuals and reduces the spread of pathogens.

Vaccines

Different **vaccines** give protection against bacterial disease such as diphtheria and tetanus, or against viral diseases such as measles, mumps and polio.

A vaccine contains a small amount of a dead or inactive form of a pathogen. → The vaccine causes white blood cells to make antibodies, in the same way they would if the body was infected by live pathogens. → If the live pathogen infects you later, your immune system remembers how to destroy it, and responds quickly so you don't fall ill. You are immune.

Depending on the particular vaccine, vaccination involves swallowing the vaccine, or having an injection into the skin or muscle.

Benefits and risks of vaccination

- Vaccinated people become immune to a disease without getting it.
- The spread of pathogens is reduced when a large proportion of the population is immunised (**herd immunity**).

- Some people develop side effects such as soreness and swelling, or show mild symptoms of the disease.
- Some vaccinations, particularly against flu, only give partial protection if there are different strains of the pathogen.

Worked example

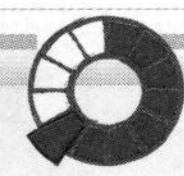

The MMR vaccine protects against measles, mumps and rubella (German measles). Children are given the vaccine when they are one year old, and again before starting school in case they did not become immune to measles the first time. The graph shows how the incidence of measles and the percentage of vaccinated children changed between 1996 and 2008.

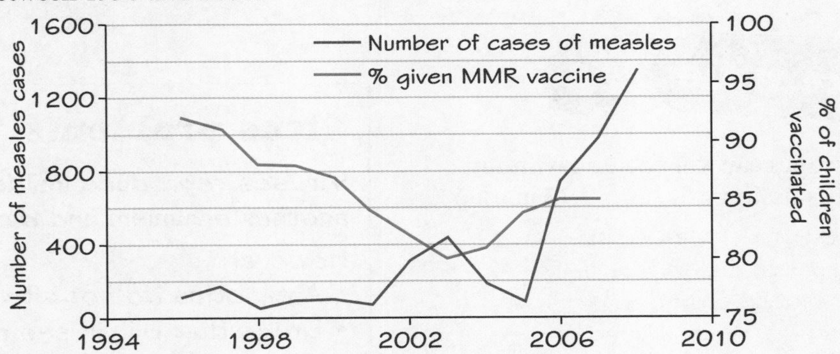

Use the graph to explain the importance of vaccination in preventing disease. **(3 marks)**

The percentage of children given the MMR vaccine decreased from over 90% in 1996 to about 80% in 2003. Although the percentage began to increase again, the number of measles cases increased rapidly after 2005. This suggests that the risk of being infected increases when the proportion of children vaccinated decreases.

Now try this

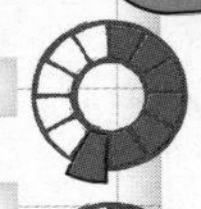

1 Describe the difference between a vaccine and vaccination. **(2 marks)**

2 In the UK, chickens and other poultry are vaccinated against *Salmonella* bacteria. Explain why this helps to control the spread of salmonella food poisoning. **(2 marks)**

Antibiotics and painkillers

Painkillers are medicines that treat **symptoms** of a disease but do not kill pathogens.

Symptoms are the results of a disease, such as aches and pains, and having a high temperature.

Antibiotics

An **antibiotic** is a medicine that kills infective bacteria inside the body:

- This helps to cure bacterial diseases.
- Specific bacteria are only killed by specific antibiotics, so doctors must prescribe the correct antibiotic.

👍 The use of antibiotics has greatly reduced deaths from infectious bacterial diseases.

👎 Strains of bacteria that are resistant to antibiotics have emerged.

Penicillin is an example of an antibiotic (you can revise the discovery of medicines on page 41).

Antibiotic resistance

Some strains of bacteria have developed a resistance to antibiotics. Certain antibiotics no longer kill these bacteria. MRSA is a strain of bacteria that has developed resistance to several antibiotics.

Mutation can produce new **antibiotic-resistant** strains of bacteria. ⇒ When the antibiotic is used, non-resistant bacteria die, but resistant bacteria survive and reproduce. ⇒ The population of resistant bacteria increases. Infections can only be treated with a new antibiotic. ⇒ If there is no new antibiotic to control the infection, it may spread rapidly, causing an epidemic or pandemic.

This is an example of natural selection.

You can revise natural selection on pages 74 and 79.

Worked example

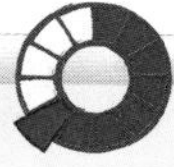

Explain how antibiotics can kill bacteria without damaging the body's tissues. **(3 marks)**

Antibiotics can prevent processes that happen in bacterial cells, but not in animal cells. For example, some antibiotics stop the bacteria forming their cell walls. This prevents them growing or dividing. Animal cells do not have cell walls, so they are not affected.

Some antibiotics are toxic to some animals, e.g. penicillin is toxic to guinea pigs.

Virus problems

Viruses reproduce inside the cells of another organism and damage these cells.

However:

- Antibiotics do not affect viruses.
- Drugs that kill viruses may also harm human cells.

It is difficult to develop antiviral drugs that do not also damage the body's tissues, so viral diseases can be difficult to treat.

Now try this

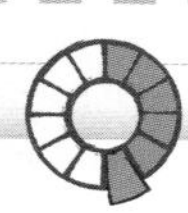

1 Explain why antibiotics are not effective against measles. **(2 marks)**

2 Give a reason why it is difficult to develop drugs that kill viruses. **(1 mark)**

What sort of pathogen causes measles?

New medicines

New medical drugs have to be tested and trialled to check that they are safe and effective.

Discovering drugs

A medical **drug** is a substance that treats the cause of a disease, or its symptoms.

Drugs were traditionally extracted from plants and microorganisms (see box on the right).

Most new drugs are **synthesised** (made using chemical reactions) by scientists working in the **pharmaceutical industry**.

Traditional sources of drugs

Drug	Origin	Treats
digitalis (digoxin)	foxglove plants	irregular heartbeat
aspirin	willow bark and leaves	aches and pains
penicillin	*Penicillium* mould	bacterial diseases

discovered by Alexander Fleming (1881–1955)

Drug testing

New drugs are tested extensively to determine their:

- **toxicity** – how poisonous or harmful they might be
- **efficacy** – how well they work at treating a disease
- **dose** – how much drug must be given for it to work.

Preclinical testing

New drugs are tested in the laboratory using cells and tissues. Promising substances are then tested on live animals.

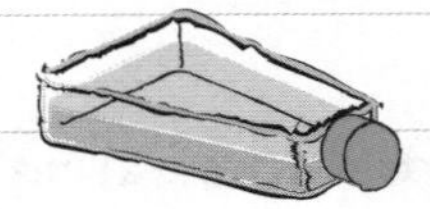

cultures of cells

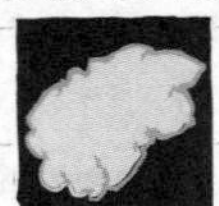

cultures of tissues

animals

2 Clinical trials

The next stages use people:

- Very low doses are given to healthy volunteers to check that the drug is not toxic to people.
- If the drug is found to be safe, different doses are given to patients to test its efficacy and **optimum dose** (the dose that works the best).

Worked example

Double blind trials may be used in clinical trials with patients. Explain what is meant by a double blind trial. **(3 marks)**

Some patients are given the drug and some are given a placebo. This is designed to look just like the drug, but it does not contain any of the drug. The doctors and patients do not know who is given the drug and who is given the placebo.

It is possible for people to feel better after taking a medicine, just because they expect to feel better. This is the **placebo effect.**

Double blind trials try to minimise this effect. The doctors and patients only find out who received the placebo after the trial. This way, differences between the two groups of patients should be due to the drug.

Now try this

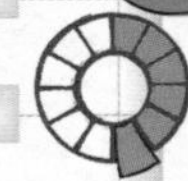

1 Describe the origin and use of one drug traditionally extracted from plants. **(2 marks)**

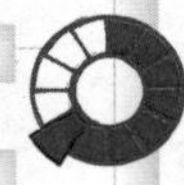

2 In a blind trial, the patients do not know who receives the placebo, but the doctors do. Suggest a reason why this type of trial is less satisfactory than a double blind trial. **(2 marks)**

Plant disease and defences

Plants have a range of adaptations to defend against pathogens and insect pests.

Examples of diseases

Disease	Pathogen
tobacco mosaic virus, TMV	virus
rose black spot	fungus
bacterial wilt	bacteria

You can revise the symptoms of TMV infection on page 34, and the symptoms of rose black spot infection on page 36.

Aphids are insect pests. Their sharp mouthparts can reach phloem, so they can feed on the sugary solution there. Aphids act as disease **vectors** if their mouthparts carry plant pathogens.

Worked example

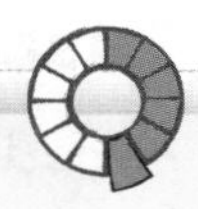

Plants can be damaged by a range of mineral ion deficiency conditions.

(a) Give two ions that are important for plant growth. **(2 marks)**

nitrate ions, NO_3^-, and magnesium ions, Mg^{2+}.

(b) Describe the importance of the ions given in part (a). **(3 marks)**

Nitrate ions are needed for protein synthesis, and magnesium ions are needed to make chlorophyll. Plants grow poorly if they cannot absorb enough of these ions.

You can revise the absorption of mineral ions by root hair cells on pages 5 and 15.

Physical defences

1. **Cellulose cell walls**
The tough **cell walls** make it difficult for microorganisms to pass through. If a part of a plant is broken or aphids penetrate the **phloem** tissue, microorganisms can more easily infect the plant.
2. **Waxy leaf cuticle**
The upper epidermis of leaves is covered by a tough waxy **cuticle**. This reduces water loss from the leaf and it also protects against the entry of microorganisms.
3. **Layers of dead cells**
Tree **bark** and layers of dead cells around stems form barriers against microorganisms. When bark falls off, the microorganisms fall off with it.

Chemical defences

Plants have chemical defence responses:
- antibacterial substances
- toxins (poisons).

Plant toxins deter **herbivores** from eating the plant. For example, a toxin in foxgloves affects the heart rate of herbivores. This toxin, digitalis, is used as a treatment for irregular heartbeat in humans (see page 41).

Mechanical adaptations

Plants have mechanical adaptations, including:
- thorns and hairs to deter animals
- leaves that drop or curl when touched.

Plants also use **mimicry** to trick animals. For example, the leaves of some species of passion flowers develop parts that look like butterfly eggs. Butterflies avoid laying eggs where there are eggs already. This means that fewer caterpillars hatch and eat the plant.

Now try this

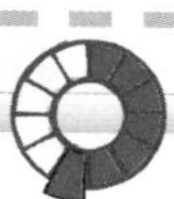

Describe **three** ways in which physical barriers can protect a plant from pests and pathogens. **(3 marks)**

Extended response – Infection and response

There will be at least one 6-mark question on your exam paper. For these questions, you will need to think scientifically and structure your answer logically, showing how the points you make are related to each other.

For the questions on this page, you can revise new medicines on pages 39 and 41.

Worked example

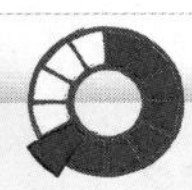

Schistosomiasis is a tropical disease caused by a parasitic worm. It leads to long-term disability, liver damage and even death. Over 60 million people each year are treated for the disease.

A single dose of a drug called praziquantel results in the death of the worms in the body. It is cheap and its side effects are mainly due to dying worms. However, praziquantel-resistant strains of the worm are appearing, so scientists are researching new drugs.

Explain how a new drug to treat the disease must be tested before it is licensed for use. **(6 marks)**

The drug must first be tested on cells, tissues and live laboratory animals. This is to work out how toxic it might be, whether it works and what a suitable dose might be.

If the drug passes the first stages, it is then tested at very low doses using healthy volunteers and patients. This is to check whether it is safe to use in people, and to find out if it has any side effects.

A drug that passes these stages is then further tested in patients to find its optimum dose. Double blind trials are used to ensure valid results and to avoid false claims of success. Patients are randomly put into one of two groups. One group receives the drug. The other group receives a placebo which does not contain the drug. The groups are only revealed at the end of the trial so the results can be analysed.

These parasitic worms are not viruses, bacteria, protists or fungi. However, you should be able to apply your knowledge and understanding to different situations using given information.

Praziquantel has desirable features, so the emergence of resistant strains will cause concern amongst health professionals.

Command word: Explain

When you are asked to **explain** something, you must include reasoning or justification of the points you make.

This part of the answer describes preclinical testing and gives reasons for doing it.

This part of the answer describes clinical trials and gives reasons for doing them.

Double blind trials are explained here. Note that, for ethical reasons, people in the placebo group would be given praziquantel afterwards to treat their infection.

Now try this

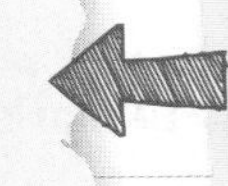

The MMR vaccine protects against measles, mumps and rubella. The graph shows the percentage of young children given the MMR vaccine in England. The World Health Organization recommends that at least 95% should be immunised. Explain the importance of this target. **(6 marks)**

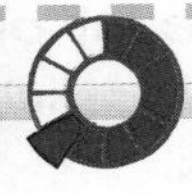

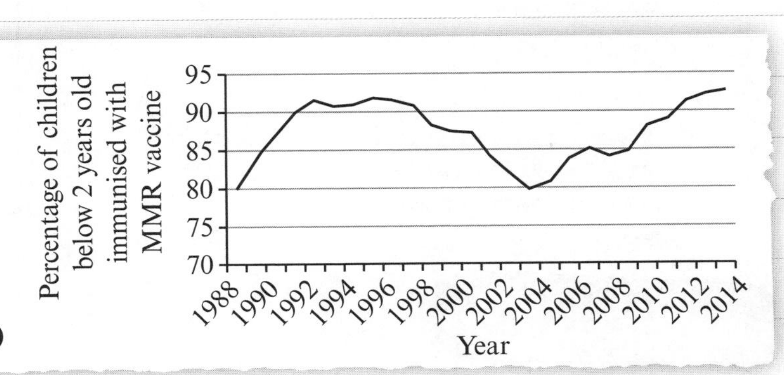

You can revise vaccination on page 39.

Photosynthesis

Plants make their own food by **photosynthesis**.

Reactants and products

Photosynthesis is an **endothermic** reaction – energy is transferred to the chloroplasts.

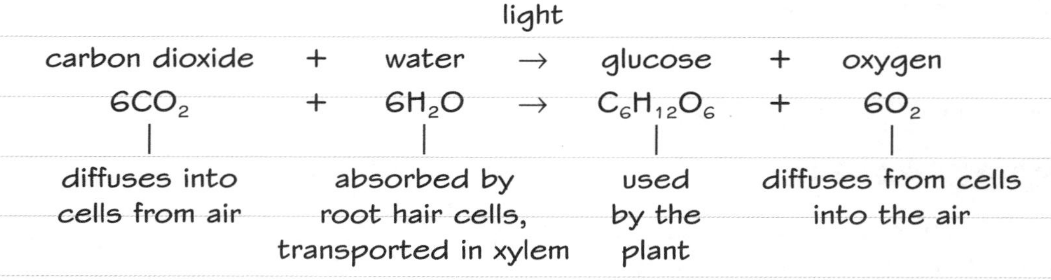

Rate of photosynthesis

The rate of photosynthesis depends upon **four** factors:

- temperature
- carbon dioxide concentration
- light intensity
- amount of chlorophyll.

The graph shows how the rate of photosynthesis varies with **temperature**.

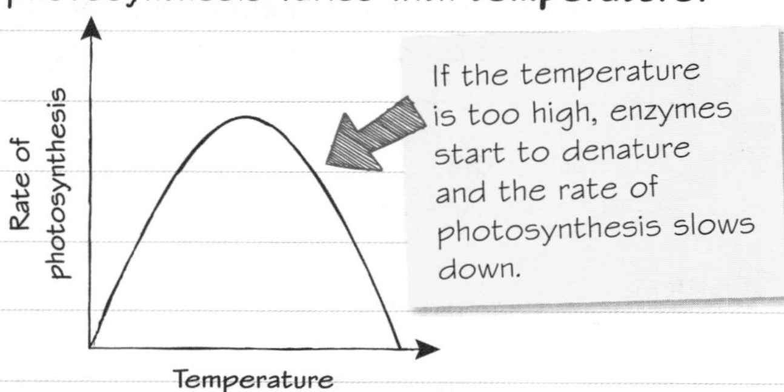

The graph shows how the rate of photosynthesis varies with **carbon dioxide concentration**. The shape of this graph:

- is different from the temperature graph
- is the same for **light intensity** and **amount of chlorophyll**.

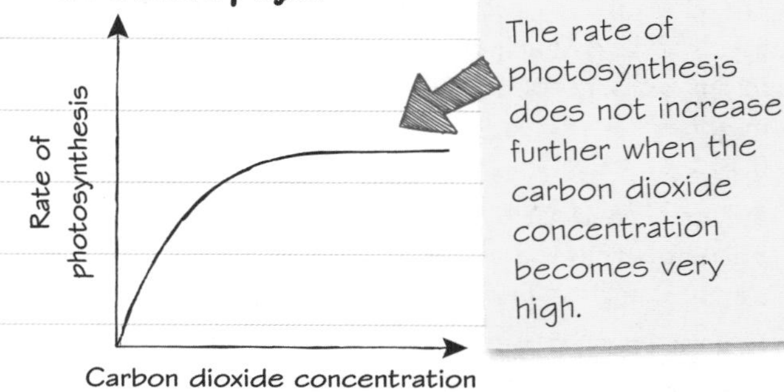

Worked example

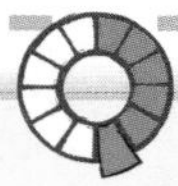

Describe how a leaf may be tested for starch. **(4 marks)**

Heat the leaf for a minute in boiling water. Transfer the leaf to a test tube of ethanol, and heat in a hot water bath for a few minutes. Rinse the leaf with cold water and spread it on a white tile. Add a few drops of iodine solution. The leaf turns blue-black where starch is present.

Boiling water kills the leaf cells and softens their cell walls so iodine solution can enter. Ethanol removes the chlorophyll.

Uses of glucose

Glucose may be converted into insoluble starch for storage. Glucose may be used:

- for **respiration**
- to produce fat or oil for storage
- to produce cellulose, which strengthens the cell wall
- to produce amino acids for proteins.

To produce proteins, plants also use nitrate ions absorbed from the soil.

Now try this

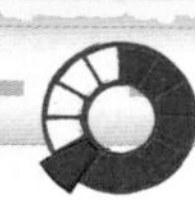

Variegated leaves have white patches or stripes inbetween green areas. A variegated leaf is starch tested as described in the Worked example. Explain why the leaf only turns blue-black where the leaf was green.

(4 marks)

Investigating photosynthesis

You can investigate the effect of **light intensity** on the **rate of photosynthesis** in an aquatic plant.

Core practical

Investigating the effect of light intensity

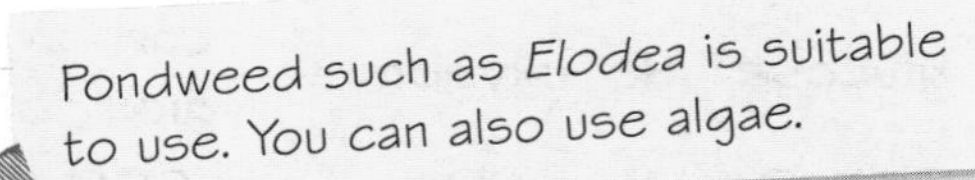

Pondweed such as *Elodea* is suitable to use. You can also use algae.

Aim

to investigate the effect of light intensity on the rate of photosynthesis in pondweed

Sodium hydrogen carbonate solution, $NaHCO_3$, is a source of carbon dioxide. It is included so that the carbon dioxide concentration is not a limiting factor.

Apparatus

- boiling tube and test-tube rack
- bright lamp, ruler, stop watch
- sodium hydrogen carbonate solution
- freshly cut piece of pondweed

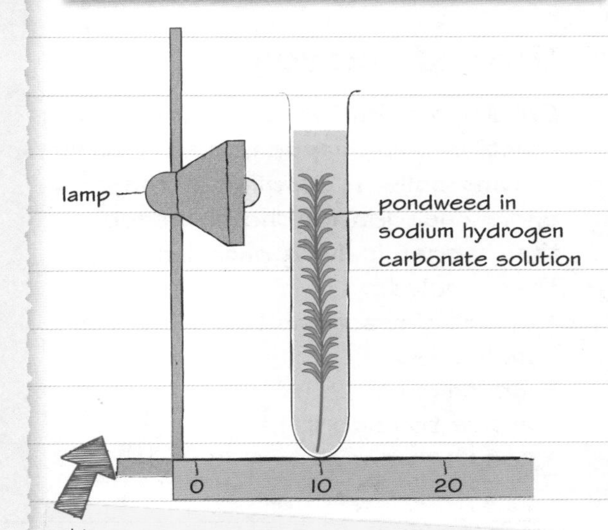

Method

1 Put the boiling tube in the test-tube rack.
2 Add sodium hydrogen carbonate solution.
3 Lower the piece of pondweed into the boiling tube.
4 Position the boiling tube 10 cm away from the lamp, and leave it for 5 minutes.
5 Count the number of bubbles produced in 1 minute. Repeat this twice more.
6 Repeat steps 4 and 5 at different distances.

Use an LED lamp to avoid heating the pondweed. If you use an ordinary lamp, place the boiling tube in a beaker of water to reduce heating.

Results

Record your results in a suitable table. This one shows some sample mean results.

Distance (cm)	Bubbles per minute			
	Run 1	Run 2	Run 3	Mean
10				50
20				35
30				30
40				27

Repeating the count at each distance allows you to check whether the results are reliable. You can do more repeats if, for example, you obtain very different results at a particular distance. The most reliable results are identical or very similar to each other.

Analysis

1 Calculate the mean rate of bubbling at each distance.
2 Plot a graph with:
- mean rate of bubbling on the vertical axis
- distance on the horizontal axis.

3 Describe how the rate of bubbling changes as the distance increases.

Choose sensible scales that allow the plotted points to occupy at least 50% of the area of the graph. Draw a curve of best fit.

Now try this

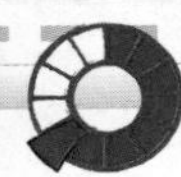

1 Draw a suitable graph to show the sample mean results in the table above. **(3 marks)**
2 Describe what the results show. **(2 marks)**

Respiration

Respiration in cells can happen **aerobically** (using oxygen) or **anaerobically** (without oxygen).

Respiration compared

Aerobic respiration	Anaerobic respiration
glucose + oxygen → carbon dioxide + water $C_6H_{12}O_6 + 6O_2 \rightarrow 6CO_2 + 6H_2O$	in muscles: glucose → lactic acid $C_6H_{12}O_6 \rightarrow 2C_3H_6O_3$
uses oxygen	does not use oxygen
complete oxidation of glucose	incomplete oxidation of glucose
exothermic reaction	exothermic reaction

Aerobic respiration transfers much more energy than anaerobic respiration.

Uses of energy

Cellular respiration is an exothermic reaction which happens continuously in living cells. It supplies all the energy needed for the processes that happen in living organisms. These include:

- chemical reactions to build larger molecules
- movement
- active transport
- keeping warm.

Fermentation

Anaerobic respiration can also happen in plant cells and yeast cells. The overall reaction is different from the reaction in muscles:

glucose → ethanol + carbon dioxide

$$C_6H_{12}O_6 \rightarrow 2C_2H_5OH + 2CO_2$$

Anaerobic respiration in yeast cells is called **fermentation**. This reaction is useful in:

- making bread – the gas makes dough rise
- making alcoholic drinks such as wine and beer.

Respiration vs photosynthesis

The overall reaction for aerobic respiration is the reverse of the overall reaction for photosynthesis. The table shows some other differences between the two processes.

Aerobic respiration	Photosynthesis
exothermic	endothermic
gives out energy	takes in light energy
uses oxygen	produces oxygen
produces carbon dioxide	uses carbon dioxide
happens in mitochondria	happens in chloroplasts
happens in animal cells and plant cells	happens in plant cells

Worked example

Compare aerobic respiration with anaerobic respiration in muscles. **(2 marks)**

Both processes use glucose. Aerobic respiration needs oxygen but anaerobic respiration does not.

Glucose is not completely oxidised in anaerobic respiration, so it releases much less energy than aerobic respiration does.

Aerobic respiration produces two products, carbon dioxide and water. Anaerobic respiration produces just one product, lactic acid.

Remember that anaerobic respiration in muscles differs from that in plants and yeast.

Now try this

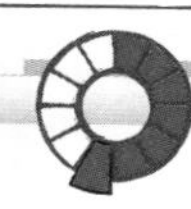

Use the information on this page to give **one** advantage and **one** disadvantage of:
(a) aerobic respiration; (b) anaerobic respiration. **(4 marks)**

Responding to exercise

The human body reacts to the increased demand for energy during exercise.

Effects of exercise

When you exercise, your muscle cells respire faster. They need more oxygen and glucose, and release more carbon dioxide. So, during exercise, the following increase:

- heart rate
- breathing rate
- breath volume.

A greater heart rate means that blood is pumped faster around the body. It can transport oxygen and glucose to cells faster, and remove carbon dioxide faster.

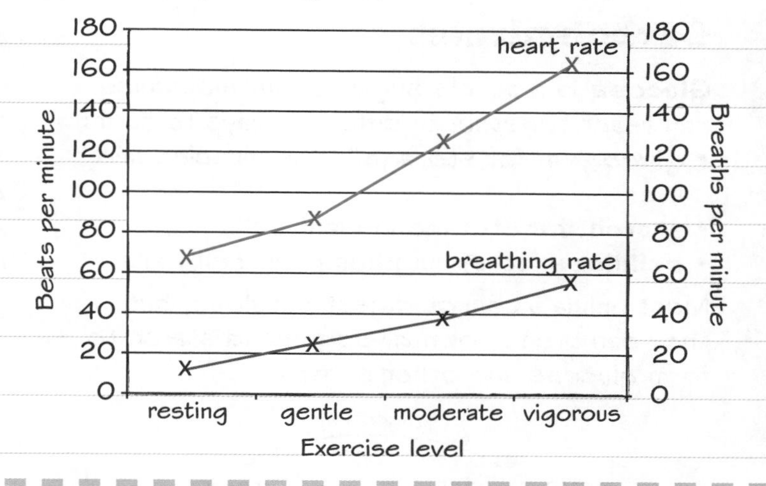

Oxygen debt

During exercise, insufficient oxygen may be supplied to the respiring muscle cells for aerobic respiration. When this happens, anaerobic respiration occurs in muscles:

- Glucose undergoes **incomplete oxidation.**
- **Lactic acid** builds up in muscles.
- The muscles become fatigued and stop contracting efficiently.

An **oxygen debt** is created, so you continue to breathe heavily for a while after exercise.

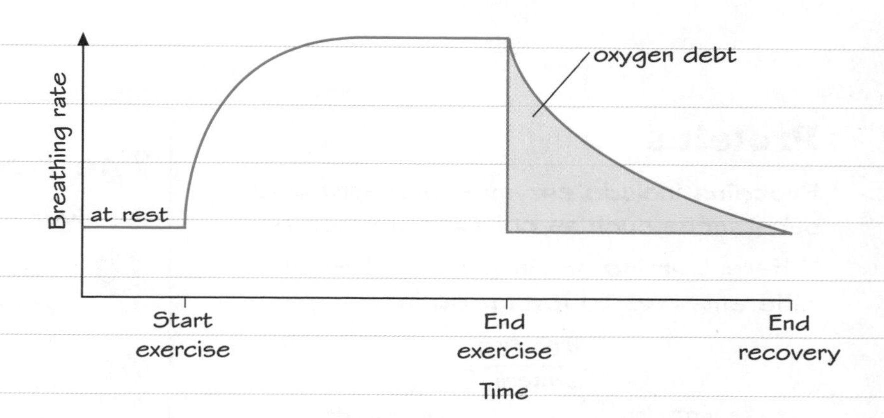

Worked example

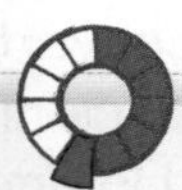

Explain why the breathing rate and breath volume increase during exercise. **(3 marks)**

A greater breathing rate means that air is inhaled and exhaled more frequently, and a greater breath volume means that more air is moved in and out each time. The rate of respiration increases during exercise, and these changes allow oxygen to be taken into the body faster and waste carbon dioxide to be released faster.

The resting breathing rate is often counted over periods of one minute. During exercise, and after it, shorter periods of 15 s or 30 s may be better.

Breath volume is measured using a device called a spirometer.

Heart rate is measured by counting the **pulse rate** in the wrist. It is usually counted during periods of 15 s, then multiplied by 4 to get the beats per minute.

Now try this

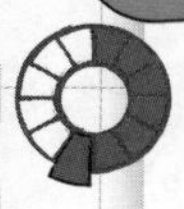

1 Explain why the muscles may become fatigued during exercise. **(2 marks)**

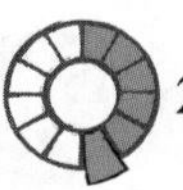

2 Give **three** changes that happen to the heart and lungs during exercise. **(3 marks)**

Metabolism

Metabolism is the sum of all the reactions that happen in a cell or in the body.

Carbohydrates

Glucose is a simple sugar. Sugar molecules can react together in different ways to form:

- **glycogen**, for storage in the muscles and liver
- **starch**, for storage in plant cells
- **cellulose**, to strengthen plant cell walls.

Most animals cannot digest cellulose, but they can break down glycogen and starch to form glucose and other sugars.

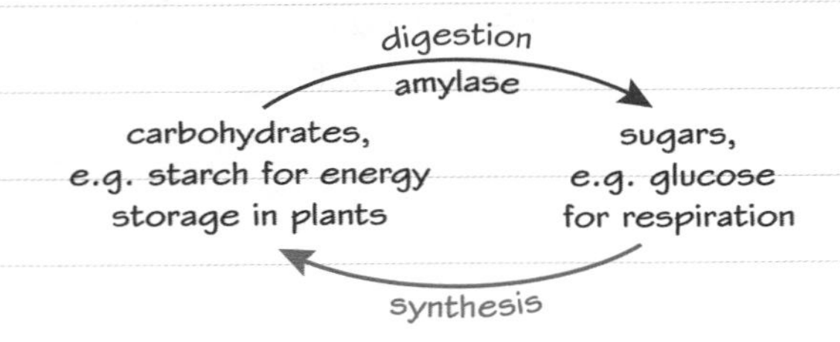

Lipids

Lipids are fats and oils. At room temperature:

- fats are solids
- oils are liquids.

A lipid molecule forms when a **glycerol** molecule reacts with three **fatty acid** molecules.

Lipids are used for:

- storage in plants and animals
- **insulation** against heat loss in animals.

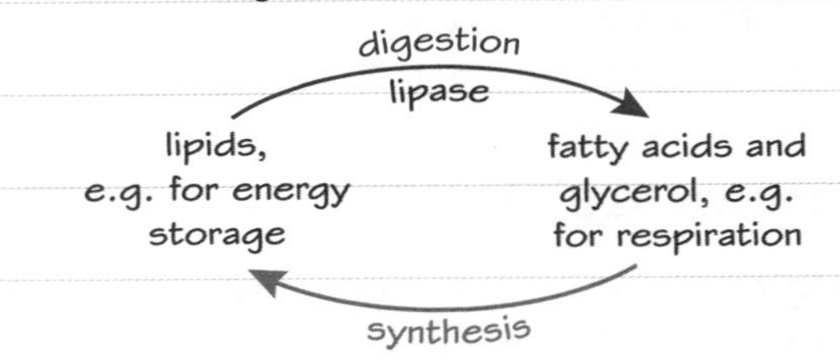

Proteins

Proteins include enzymes, and structural substances such as collagen and keratin.

Different **amino acids** join together in different ways to form proteins.

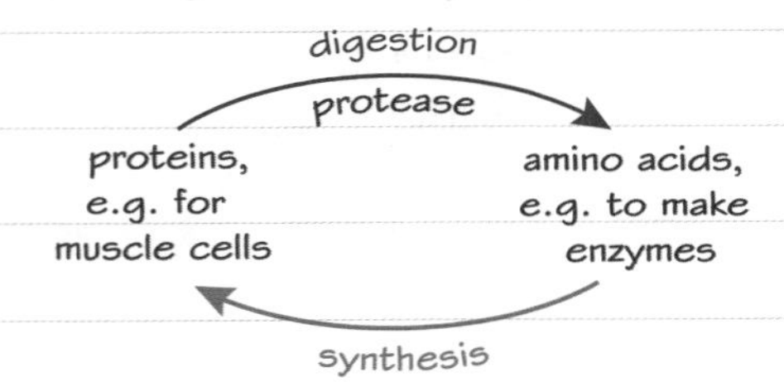

The liver

The **liver** has many functions. These include:

1. converting lactic acid, produced during anaerobic respiration, into glucose
You can revise this on page 46.

2. converting excess glucose into glycogen for storage
You can revise this on page 58.

3. converting excess amino acids to form ammonia **(deamination)**, and converting this ammonia into urea for excretion.

Worked example

Describe how plants form proteins, starting with glucose produced by photosynthesis. **(3 marks)**

Plants absorb nitrate ions through their roots. Glucose and nitrate ions combine to form amino acids. These join end to end to form proteins.

Remember that energy transferred by respiration is used by organisms for continual processes, controlled by enzymes, that synthesise new substances.

The details of how amino acids are formed are complicated and involve many reactions.

Now try this

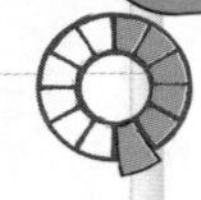

1 Give the meaning of 'metabolism'. **(1 mark)**

2 Name **three** types of digestive enzymes involved in metabolism. **(3 marks)**

Extended response – Bioenergetics

There will be one at least one 6-mark question on your exam paper. For these questions, you will need to think scientifically and structure your answer logically, showing how the points you make are related to each other.

For the questions on this page, you can revise responding to exercise on page 47.

Worked example

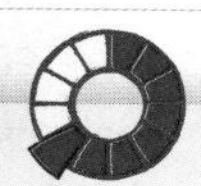

Explain the effects of temperature, light intensity, carbon dioxide concentration and the amount of chlorophyll on the rate of photosynthesis. **(6 marks)**

Photosynthesis involves the reaction between carbon dioxide and water:

$$\text{carbon dioxide} + \text{water} \xrightarrow{\text{light}} \text{glucose} + \text{oxygen}$$

As carbon dioxide is one of the reactants, its concentration in the air affects the rate of photosynthesis.

As the concentration increases, the rate of photosynthesis increases until it reaches a maximum rate determined by another factor.

Rate of photosynthesis
Carbon dioxide concentration

Photosynthesis is an endothermic reaction in which energy is transferred to the chloroplasts by light. So the greater the light intensity, the greater the rate of photosynthesis. Chlorophyll in chloroplasts absorbs light, so the more chlorophyll there is, the greater the rate of photosynthesis. Graphs of rate against light intensity or amount of chlorophyll are similar in shape to the one sketched above.

The rate of chemical reactions increases as the temperature increases. So the rate of photosynthesis increases as the temperature increases. However, enzymes are involved in photosynthesis. They become denatured at high temperatures, so the rate of photosynthesis decreases above a certain temperature.

Command word: Explain

When you are asked to **explain** something, you must include reasoning or justification of the points you make.

Showing the equation helps to explain why carbon dioxide and light are important in photosynthesis.

If you sketch a graph, it may help save some writing and make your answer clearer.

This part of the answer links the absorption of light to the rate of photosynthesis. There is no need to sketch a graph since one is already given, and its relevance is explained here.

The answer could include that an enzyme becomes denatured when the shape of its active site changes irreversibly.

Now try this

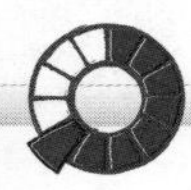

Explain why breathing rate and heart rate increase during exercise. **(6 marks)**

Include aerobic respiration and anaerobic respiration in your answer (you can revise these on pages 46 and 47).

Homeostasis

Homeostasis involves the regulation of the internal conditions of a cell or organism.

Optimal conditions

Enzymes become **denatured** and no longer work if the temperature becomes too high, or if the pH becomes too high or low. The **optimum temperature** and **optimum pH** are the conditions in which an enzyme works best.

Homeostasis:
- maintains optimal conditions for enzyme action and all cell functions
- involves responses to changes in the internal or external conditions.

Control in the body

Many conditions are controlled in the human body. They include:

1. body temperature
   You can revise thermoregulation on page 56.
2. blood glucose concentration
   You can revise this on page 58.
3. water levels
   You can revise osmoregulation on page 60.

Control systems

Homeostasis uses automatic control systems. These may include:
- nervous responses (involving the nervous system)
- chemical responses (involving hormones).

All control systems have three main parts.

The diagram below outlines how these work to control the amount of water in the body.

1. **Receptors**. These are cells that detect **stimuli** (changes in the environment).
2. **Coordination centres**, such as the brain, spinal cord and pancreas. These receive information from receptors and process it.
3. **Effectors**, such as muscles or glands. These bring about the responses needed.

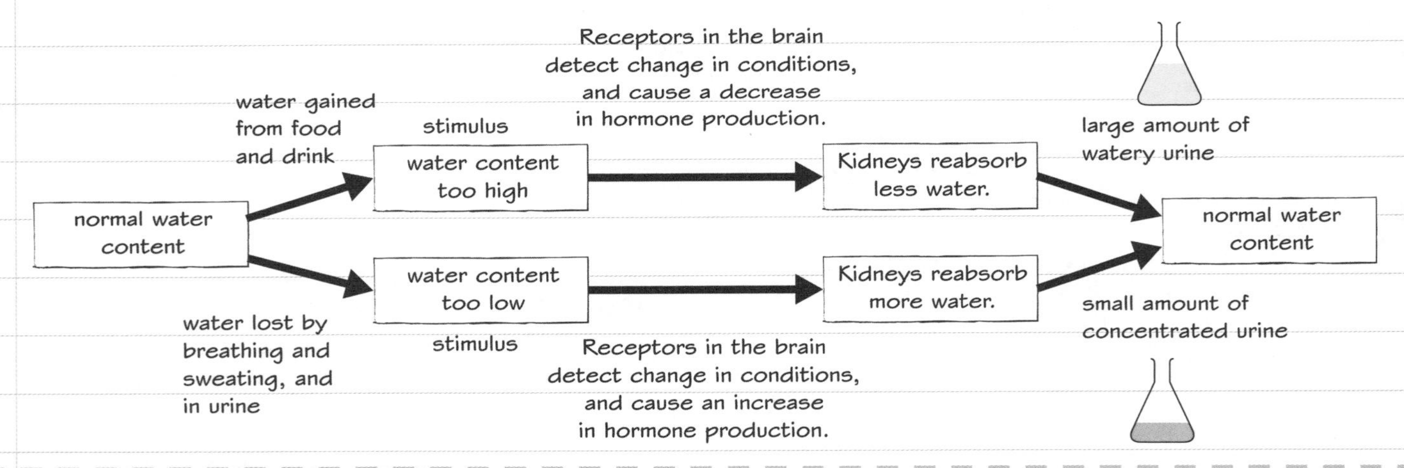

Worked example

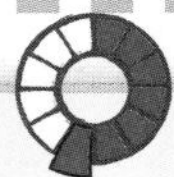

Describe why the control mechanisms in homeostasis may be described as 'negative feedback mechanisms'. **(2 marks)**

When a change happens, the control mechanisms act to reverse the change. For example, if the water content increases, the control mechanism decreases it again.

If the water content decreases, the control mechanism increases it again.

Homeostasis also works to bring body temperature and blood glucose levels up or down to their optimal values.

Now try this

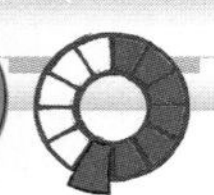

Explain what is meant by 'homeostasis'. **(3 marks)**

Neurones and the brain

The **nervous system** allows us to react to our surroundings and to coordinate our behaviour. It contains cells called **neurones**. The **central nervous system** (CNS) is the brain and spinal cord.

Types of neurones

There are three main types of neurones:

- **Sensory neurones** carry electrical impulses from **receptors** to the CNS.
- **Relay neurones**, found in the **CNS**, carry impulses from sensory neurones to motor neurones.
- **Motor neurones** carry impulses from the CNS to **effectors**, such as muscles and glands.

Sensory neurone

The electrical impulse jumps from one gap in the myelin sheath to the next, speeding up the rate of transmission.

The structure of motor neurones is shown on page 66.

skin receptor cells

direction of impulse

cell body

Axon terminals pass impulses to other neurones.

dendron

axon

fatty **myelin sheath**

Dendrites collect impulses from receptor cells.

carry impulses long distances

This insulates the neurone. The electrical impulse cannot cross the fatty myelin sheath.

Relay neurone

Dendrites collect impulses from sensory neurones.

cell body

axon

Axon terminals pass impulses to other neurones.

Brain structure

The **brain** is made of billions of interconnected neurones and controls complex behaviour.

The cerebral cortex controls **voluntary movement**, **interprets** sensory information and is responsible for **learning** and **memory**.

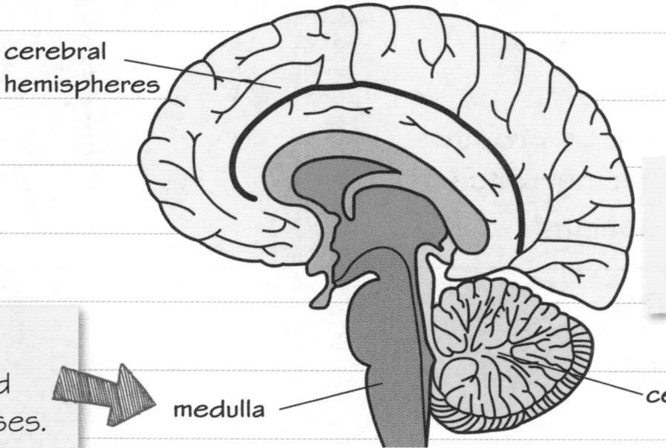

The cerebellum coordinates and controls precise and smooth **movement**.

The medulla regulates the **heartbeat**, **breathing** and other unconscious processes.

Worked example

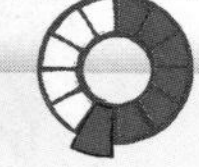

The cerebellum is also likely to be involved during exercise, coordinating and controlling precise movements.

Name the part of the brain responsible for:

(a) recognising a person from a photograph **(1 mark)**

the cerebral cortex

(b) increasing heart rate during exercise. **(1 mark)**

the medulla

Now try this

1 Compare the roles of sensory, relay and motor neurones in the nervous system. **(3 marks)**

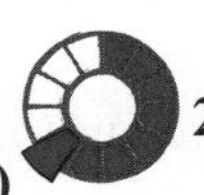

2 Explain how the structure of a sensory neurone is related to its function. **(4 marks)**

Reflex actions

The **central nervous system**, CNS, is the brain and spinal cord.

Coordinating responses

The CNS coordinates the response of effectors:

Reflex actions do not involve the conscious part of the brain. This means that they are:

- automatic (you do not have to think for them to happen)
- rapid.

Reflex actions are **innate** – they are not learned responses.

They are important because they:

- control basic functions such as breathing
- help to avoid danger and harm.

A **reflex arc** works like this:

Synapses

The point where two neurones meet is a **synapse**. Electrical nerve impulses cannot cross the gap in a synapse, so the signal is transmitted by **neurotransmitters**. This process:

- works in one direction only
- allows one nerve impulse to produce impulses in several other neurones
- is slower than nerve impulses.

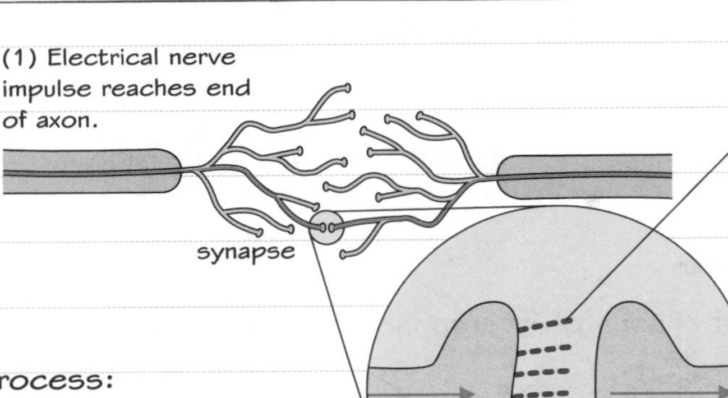

Worked example

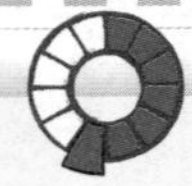

The diagram shows structures involved in a reflex arc.

(a) Name the structures labelled A, B and C in the diagram. **(3 marks)**

A – sensory neurone B – relay neurone

C – motor neurone

(b) Explain how the reflex shown helps to increase the chance of survival. **(2 marks)**

A hot flame or surface would damage the skin. The reflex allows the hand to move away quickly enough, without conscious thought, to reduce the chance of being injured.

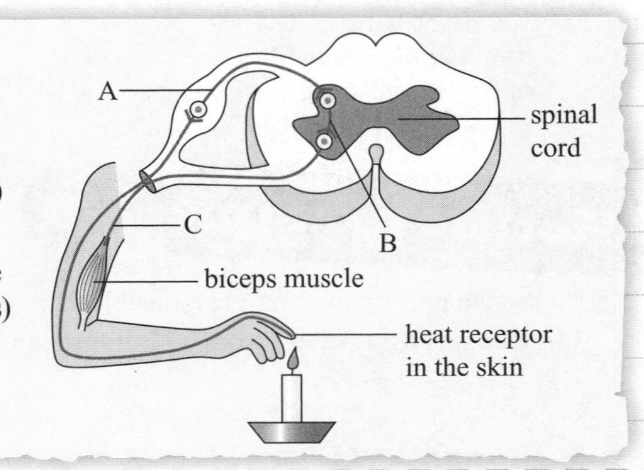

Now try this

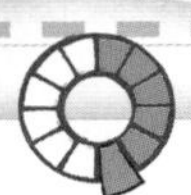

Give a reason that explains why reflex actions are automatic and rapid. **(1 mark)**

Investigating reaction times

Practical skills You can investigate the effect of a factor on human **reaction time.**

Core practical

Effect of practice on reaction times

Aim

to investigate the effect of practice on human reaction times

You could also plan and carry out an investigation to see if exercise or having a cola drink containing caffeine has an effect.

Apparatus

- ruler, 1 m long, marked in cm
- a partner to work with

You may be given a ruler with a paper scale marked in reaction times, or a table of data so you can convert readings in cm into reaction times.

Method

1. Sit on a stool with the forearm of your weaker hand on the bench. Your hand should hang over the edge.
2. Open your thumb and forefinger. The ruler will fall between them and you will need to catch it with your thumb and forefinger.
3. Your partner should then hold the ruler upright so the 0 cm mark is level with the top of your thumb.
4. Without telling you, your partner should let go of the ruler. Catch the ruler as soon as you can, using your thumb and forefinger.
5. Record the reading on the ruler, level with the top of your thumb, to the nearest cm.
6. Repeat steps 2–5 ten times.
7. Swap roles and repeat steps 1–6.

If you are right handed, your weaker hand is your left hand, and vice versa.

Sit up straight, and focus across the lab rather than on the ruler itself.

The ruler should lightly touch your thumb so you can feel when it starts to drop.

Results

Record your results in a suitable table.

Drop number	Reading (cm)	Reaction time (ms)
1		
2		

Analysis

1. For each reading, determine the equivalent reaction time using a data table or by calculation.
2. Plot a bar chart with:
   - reaction time on the vertical axis
   - drop number on the horizontal axis.
3. Describe what the results show.

Maths skills

Calculating reaction time

This ruler falls at a constant acceleration because of gravity. This means you can convert the reading in cm into time in milliseconds, ms:

reaction time in ms $= \sqrt{2000 \times \text{cm}}$

For example, if the ruler falls 75 cm:

reaction time $= \sqrt{2000 \times 75}$

$= \sqrt{150000}$

$= 387$ ms

Most readings will be to 2 significant figures, so you only need to give the final answer to 2 significant figures. In this example, 390 ms.

Now try this

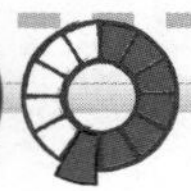

Explain why you should not do any practice before starting the investigation. **(2 marks)**

The eye

The **eye** is a **sense organ** that contains receptors sensitive to light intensity and colour.

Structure of the eye

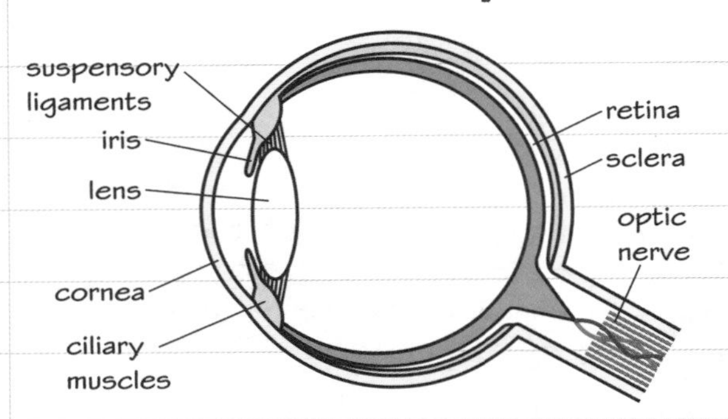

Light travels into the eye through:
- the tough, transparent **cornea**
- the pupil (the hole made by the **iris**)
- the transparent **lens**.

Light reaches the receptor cells in the **retina**:
- **Rods** are sensitive to low light levels and respond to how light or dark an object is.
- **Cones** are sensitive to bright light and respond to different colours.

Structure	Features	Function
sclera	tough, white outer layer	• protects against damage • provides attachment for eye muscles
retina	contains receptor cells (rods and cones)	• detects light • sends impulses to the optic nerve
optic nerve	contains sensory neurones	• sends impulses to the optical centre in the brain

Focusing light

Accommodation is the process of changing the shape of the lens to focus on objects.

	Focusing on a near object	Focusing on a distant object
Light rays	rays of light from a near object eye lens	rays of light from a distant object eye lens
Ciliary muscles	contract	relax
Suspensory ligaments	loosen	pulled tight
Lens	thicker refracts light strongly	pulled thin refracts light slightly

Worked example

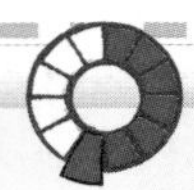

The diagrams show how the size of the pupil changes when the light becomes dimmer. Describe how this change occurs. **(3 marks)**

The iris contains muscles. In dim light, its circular muscles relax and its radial muscles contract. The pupil dilates (becomes larger). This lets more light into the eye.

bright light   dim light

This is an example of a reflex action. You can revise these on page 52.

Now try this

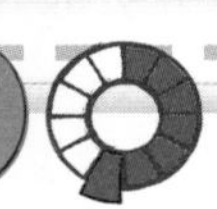

Describe the change to the pupil when we go into bright light, and how it occurs. **(3 marks)**

Eye defects

Myopia (short sightedness) and **hyperopia** (long sightedness) are common eye defects.

Myopia

A person with **short sightedness**:
- can see near objects clearly
- but distant objects appear blurred.

Light focuses in front of the retina because:
- the eyeball is too long, or
- the cornea is too curved.

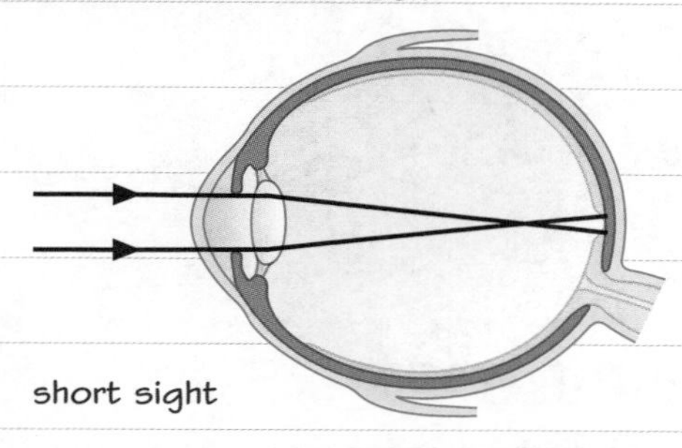
short sight

This defect is usually treated with spectacles that have **concave lenses**. These **refract** light so it **diverges** before reaching the eye. This means that light focuses onto the retina.

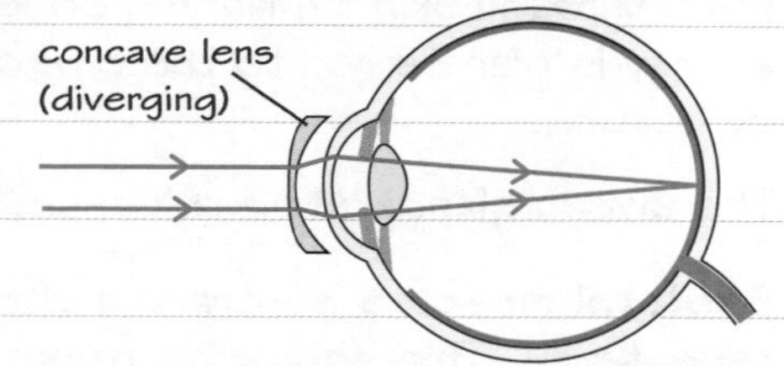

Hyperopia

A person with **long sightedness**:
- can see distant objects clearly
- but near objects appear blurred.

Light focuses behind the retina because:
- the eyeball is too short, or
- the cornea is too flat.

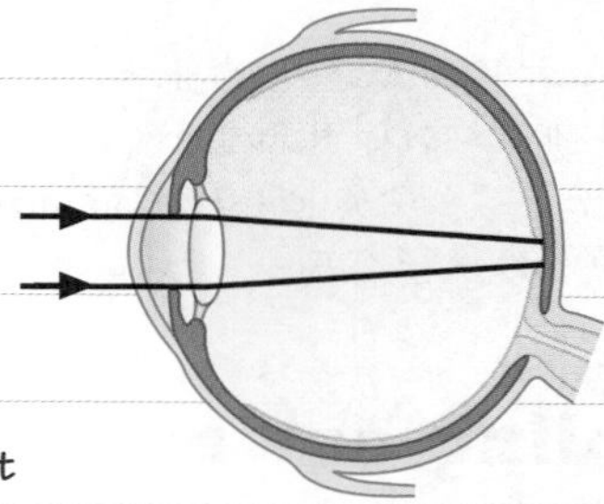
long sight

This defect is usually treated with spectacles that have **convex lenses**. These refract light so it **converges** before reaching the eye. This means that light focuses onto the retina.

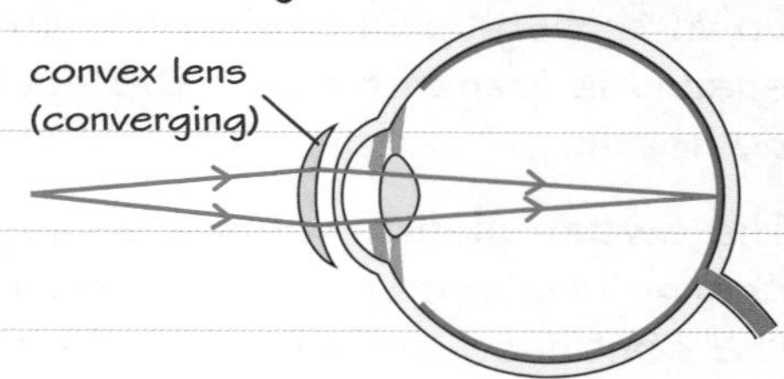

Other treatments

Other treatments for myopia or hyperopia include:
- hard or soft **contact lenses**
- **laser eye surgery**, in which a laser is used to change the cornea's shape so it refracts light differently before reaching the lens
- surgically implanting an **artificial lens**, either as a replacement for the natural lens or to work with it.

Cataracts

A **cataract** occurs when a part of the lens becomes less transparent. This stops some of the light reaching the retina, so vision becomes cloudy or blurred. Cataracts are treated by surgery:
- The natural lens is removed.
- A polymer lens is inserted.

The artificial lens has a fixed shape, so spectacles may still be needed afterwards.

Worked example

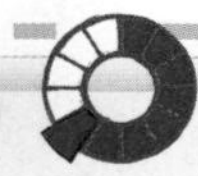

Explain why the incidence of hyperopia tends to increase with age. **(3 marks)**

The lens becomes harder as people age. This makes it increasingly difficult for the ciliary muscles to change the lens shape sufficiently. Eventually the muscle cannot make the lens thick enough to focus on near objects.

Now try this

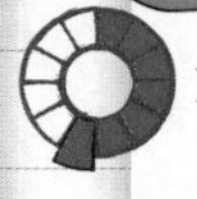

1 Compare the difference in vision between a person with myopia and a person with hyperopia. **(2 marks)**

2 Explain how spectacles are used to treat hyperopia. **(5 marks)**

Thermoregulation

Thermoregulation involves several processes that keep body temperature close to 37 °C.

The skin

The **skin** is an organ that plays a major role in thermoregulation.

releases sweat when warm to lose heat by evaporation

Vasodilation/vasoconstriction changes blood flow through surface capillaries depending on temperature.

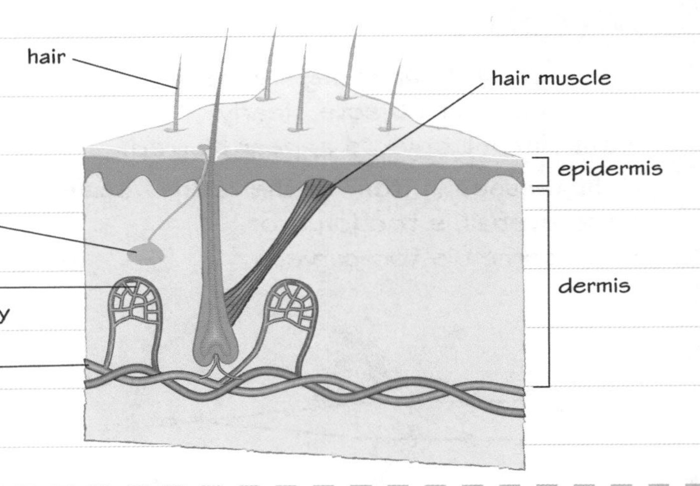

Cooling down

If the body temperature is too high:

1. **Vasodilation** occurs.
Blood vessels supplying the skin become wider (they **dilate**). More blood flows through skin capillaries, so more energy is transferred to the environment by heating.
2. The **sweat glands** release sweat.
Energy must be transferred to allow the continued evaporation of water in **sweat**. This causes a transfer of energy from the skin, cooling the body.

Staying warm

If the body temperature is too low:

1. **Vasoconstriction** occurs.
Blood vessels supplying the skin become narrower (they **constrict**). Less blood flows through skin capillaries, so less energy is transferred to the environment by heating.
2. The sweat glands stop releasing sweat.
3. **Skeletal muscles** contract and relax repeatedly. This **shivering** releases energy by heating because of increased respiration in the muscle cells.

Monitoring and control

The **thermoregulatory centre** in the brain monitors and controls body temperature.

Thermoregulatory centre contains receptors sensitive to blood temperature.

nervous impulses sent to brain

temperature receptors in skin

Worked example

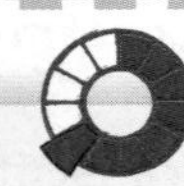

Explain how the hairs on the skin play a role in regulating body temperature. **(3 marks)**

If the body temperature becomes too low, the hair muscles contract, pulling the skin hairs upright. This traps a thicker layer of insulating air close to the skin, which helps to reduce heat losses from the skin.

This is why you get 'goosebumps' in the cold. The skin hairs lie back down again in warm conditions.

Now try this

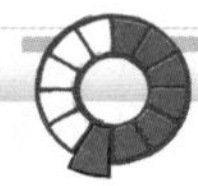

Describe **two** changes that happen if we get too hot, so that more energy is transferred from the skin to the surroundings. **(2 marks)**

Hormones

The **endocrine system** comprises glands that secrete hormones directly into the bloodstream.

Secreting hormones

A **hormone** is a substance that:

- is secreted by a **gland**
- is transported in the bloodstream
- has an effect on a **target organ**.

The **pituitary gland** is a 'master gland'. It secretes several hormones in response to body conditions. These hormones then act on other glands, stimulating the release of other hormones.

Hormones and nerves

The endocrine system and the nervous system are both involved in responses to changes in the environment. However, there are differences in their effects and how quickly they work.

	Hormones	Nerve impulses
Speed	slow	fast
Duration	long	short

Glands and hormones

Different endocrine glands produce different hormones.

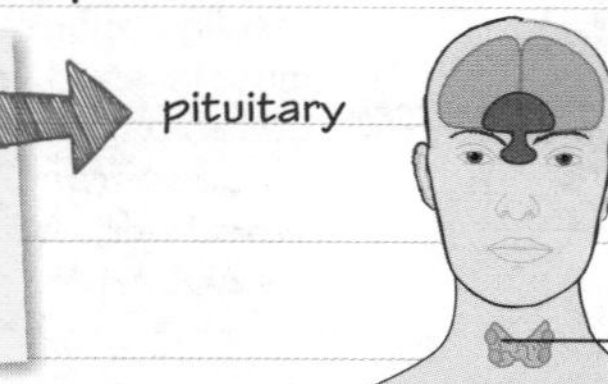

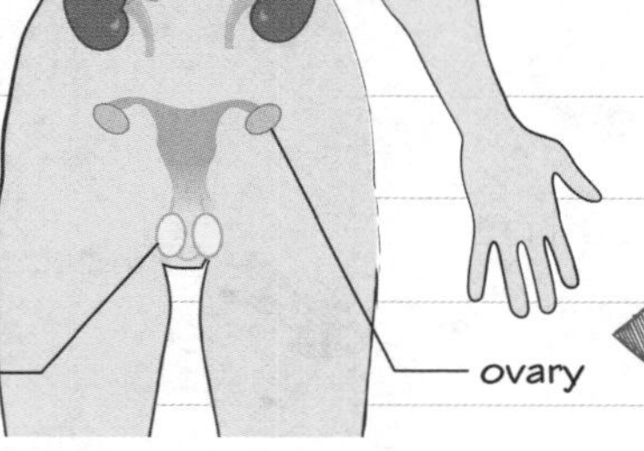

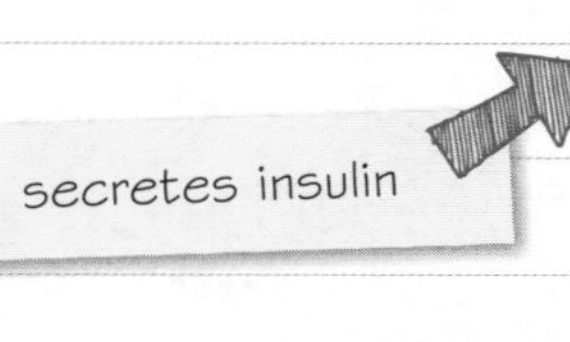

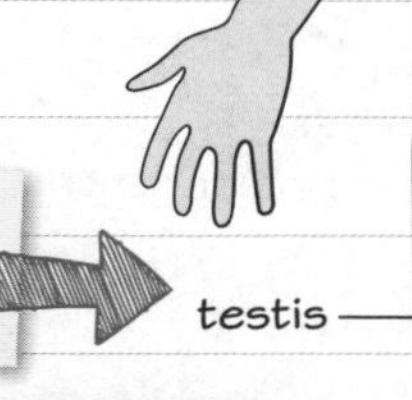

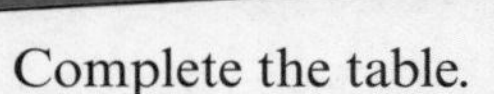

Worked example

Complete the table. **(5 marks)**

Hormone	Target organ(s)
FSH and LH	ovaries
insulin	liver, muscle
oestrogen	ovaries, uterus, pituitary gland
progesterone	uterus
testosterone	male reproductive organs

Oestrogen and testosterone are the main reproductive hormones (see page 62).

Now try this

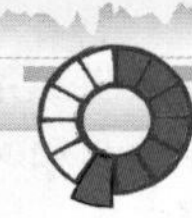

1. Describe what a hormone is. **(3 marks)**
2. Give **two** differences between hormonal responses and nervous responses. **(2 marks)**

Blood glucose regulation

The **pancreas** monitors and controls blood glucose concentration.

Insulin

Insulin is a hormone secreted by the **pancreas**. Insulin causes:

- cells to absorb more glucose, so
- blood glucose levels fall.

Muscle cells and the liver can convert the absorbed glucose into **glycogen** for storage.

Glycogen and starch

Glycogen and starch are complex molecules produced when many **glucose** molecules join together in different ways.

- Plants store starch in their cells.
- Animals store glycogen in their liver cells and muscle cells.

Controlling glucose levels

Blood glucose concentration is maintained within a small range:

- A change in concentration causes responses that reverse the change.

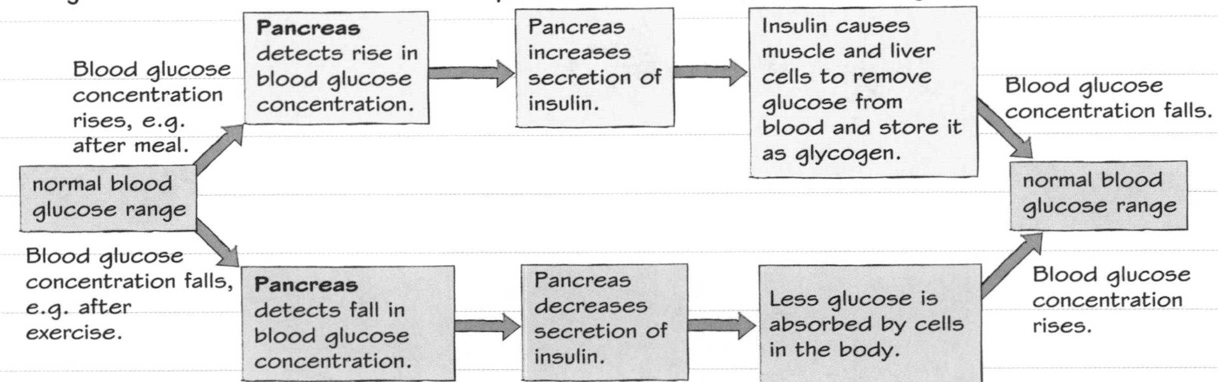

Worked example

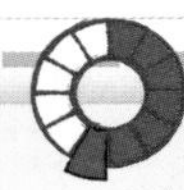

The graph shows how a person's blood glucose concentration changes after a meal.

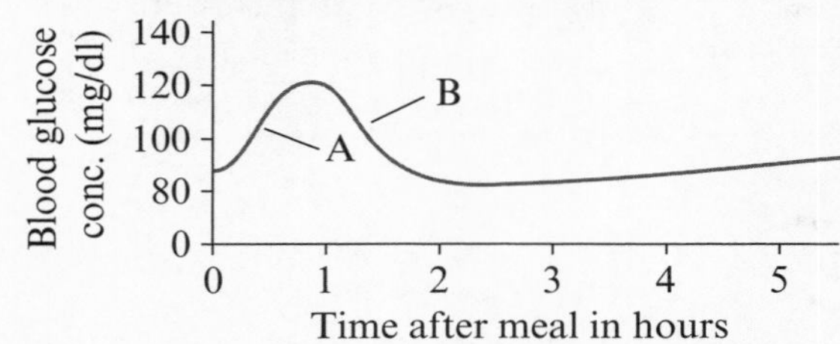

Explain the changes observed at the points marked A and B on the graph. **(5 marks)**

Remember to refer to the graph.

At A, the person has just had their meal. Starch and complex sugars are being digested to form glucose, which is absorbed into the blood. This causes the blood glucose concentration to increase over the first hour after eating.

At B, the secretion of insulin by the pancreas has increased. This causes liver and muscle cells to increase their absorption of glucose. Glucose is converted to glycogen for storage, and blood glucose concentration falls.

Now try this

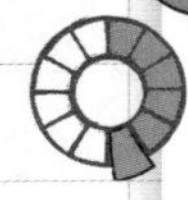

1 (a) Name the main target organ of insulin. **(1 mark)**

(b) Give **two** effects of insulin on liver cells. **(2 marks)**

2 Describe the role of the pancreas in controlling blood glucose concentration. **(2 marks)**

Look at the diagram to help you.

Diabetes

Diabetes is a condition in which the body cannot properly control blood glucose concentration.

Two types of diabetes

There are two types of diabetes, with different causes and treatments. Type 2 diabetes is associated with excess body mass, but type 1 diabetes is not.

	Type 1 diabetes	Type 2 diabetes
Cause	• Pancreatic cells are destroyed by the body's immune system. • The pancreas does not produce sufficient insulin.	• Liver cells and muscle cells no longer respond to insulin. • The pancreas may not produce enough insulin.
Treatment	insulin injections	a carbohydrate-controlled diet and exercise
Diagnosis	often in childhood	usually when over 30 years old

Body mass index

Obesity is a risk factor for type 2 diabetes. **Body mass index**, BMI, is one indicator of obesity. The table shows categories of BMI that are often used.

BMI (kg/m^2)	Category
<18.5	underweight
18.5–25	normal weight
25–30	overweight
>30	obese

Maths skills: Calculating BMI

You can calculate a person's BMI using:

$$\text{BMI} = \frac{\text{mass in kg}}{(\text{height in m})^2}$$

For example, the BMI of a 1.73 m tall person of mass 66 kg is:

$$\frac{66}{(1.73)^2} = \frac{66}{3.0} = 22\,kg/m^2$$

This person is in the normal weight category.

Worked example

The incidence of type 2 diabetes was studied in 51 529 men and 114 281women with different BMIs.

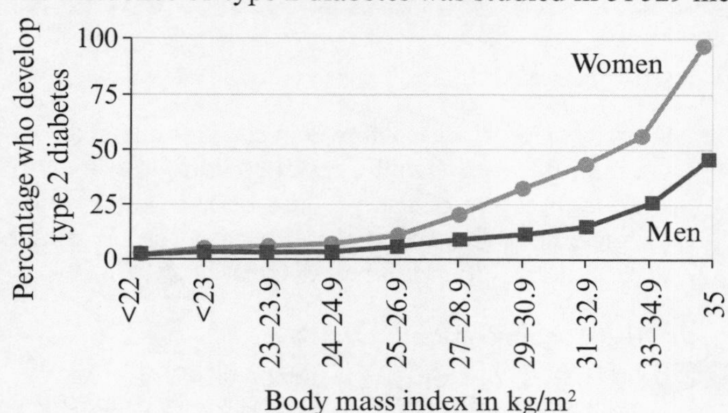

(a) Describe what the results show. **(4 marks)**

The risk of developing type 2 diabetes increases as BMI increases. The risk increases slightly in the normal weight categories (up to 24.9) but increases greatly in the obese categories (over 30). Women are more likely to have the condition than men. For example, twice as many women with the highest BMIs are affected than men.

(b) Describe the significance of the sample sizes in this study. **(2 marks)**

These are very large sample sizes, so the results are likely to be valid.

The people should also be matched for factors such as age to obtain valid results.

Now try this

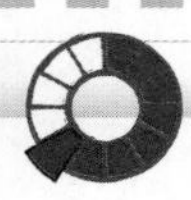

Explain how the dose of insulin needed by a person with type 1 diabetes will vary, depending on the food eaten and amount of exercise taken. **(3 marks)**

Controlling water balance

The **urinary system** maintains water balance, and removes waste products.

Water loss

Body cells do not work efficiently if they lose or gain too much water by **osmosis**:

- Water leaves the body from the **lungs** when we exhale.
- Water, ions and urea are lost from the body in **sweat.**

The body cannot control the loss of water, ions or urea by the lungs or the skin.

Control is achieved by the **kidneys** in the urinary system. Kidneys produce urine by:

- filtration of the blood
- **selective reabsorption** of water and other useful substances.

Osmotic changes and cells

Cells are affected by changes in concentration of body fluids. The effects on red blood cells are easily observed with a light microscope.

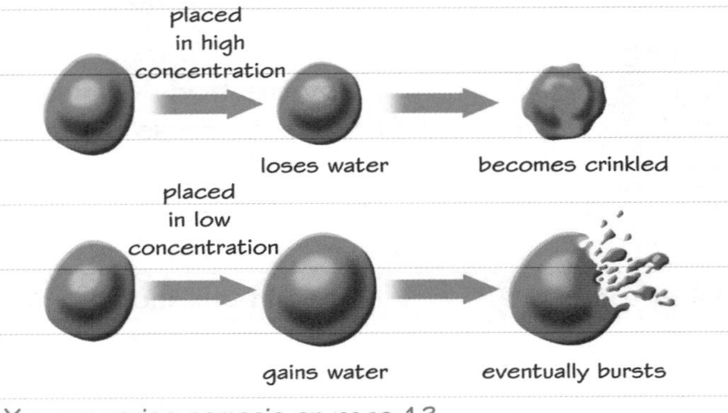

You can revise osmosis on page 13.

Structure of the urinary system

Make sure you know the difference between urea and urine, and between ureter and urethra.

Urea is produced from the breakdown of excess amino acids in the **liver**. It is toxic in excess.

The **renal veins** carry cleaned blood back to the body.

The **renal arteries** carry blood from the body to the kidneys.

The **ureters** carry **urine** from the kidneys to the bladder.

The **kidneys** remove substances including urea from the blood and make urine.

The bladder stores urine.

A muscle keeps the exit from the bladder closed until we decide to urinate.

Urine flows through the **urethra** to the outside of the body.

Worked example

The table shows the typical concentrations of substances in blood plasma and urine.

	Concentration (g/dm³)	
Substance	**Plasma**	**Urine**
urea	0.3	18
salts	9	20
glucose	1	0

Describe the differences between the composition of plasma and urine. **(4 marks)**

The concentration of urea in the urine is sixty times greater than its concentration in plasma. The concentration of salts is just over twice as much in the urine. No glucose is found in the urine, even though it is found in the plasma.

The concentrations in urine vary, depending upon how much water is reabsorbed.

After filtering the plasma, the kidneys reabsorb all the glucose and some of the salts and urea.

Now try this

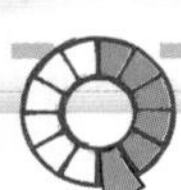

Give **two** ways in which water, ions and urea can leave the body. **(2 marks)**

Kidney treatments

People with **kidney failure** may be treated by organ transplant or by using kidney dialysis.

Organ transplant

A healthy organ may be provided by a donor. It is **transplanted** by surgery into a patient.

A healthy kidney is connected to the blood circulation, to do the work of the diseased kidneys.

Problem: The **antigens** on the transplanted kidney cells are different from antigens on cells in the patient's body.

The **antibodies** in the patient's immune system attack the transplanted kidney and **reject it**.

To prevent rejection:
- the antigens on the transplanted kidney and patient's tissues must be as similar in type as possible
- the patient must be treated for life with drugs to reduce the effects of the immune system.

This means the patient may get more infections than normal.

Antigens on the surface of cells are proteins and complex sugars that differ from person to person. You do not normally form antibodies against your own antigens.

Kidney dialysis

Kidney **dialysis** uses a machine to carry out the functions of the kidneys.

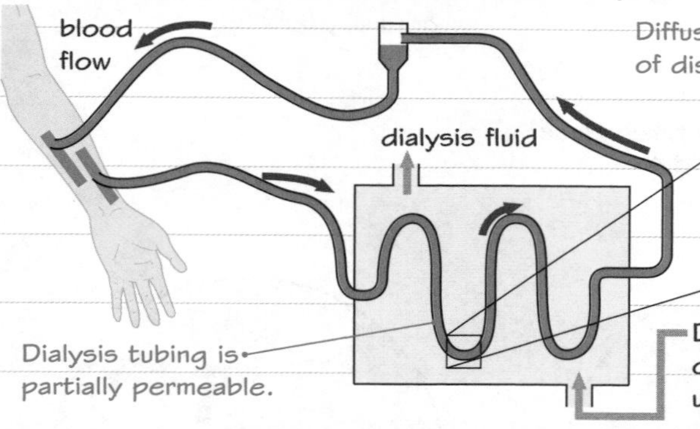

Feature	Reason
Dialysis tubing is partially permeable.	Blood cells, platelets and plasma proteins are not removed from the blood.
Dialysis fluid entering the machine has no urea.	This provides a large concentration difference so urea diffuses quickly from the blood.

Worked example

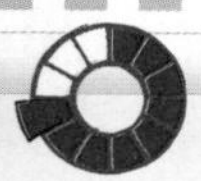

Compare the advantages and disadvantages of kidney dialysis and organ transplant as treatments for kidney failure. **(4 marks)**

A kidney transplant allows the patient to have a normal diet, free of visiting the hospital several times a week. However, donor kidneys are less readily available than dialysis machines, and carry a risk of organ rejection.

Dialysis patients must control the amount of protein in their diet to reduce the production of urea. A dialysis session takes a few hours.

Immunosuppressant drugs must be taken for life, to reduce the risk of the patient's immune system rejecting the transplanted kidney.

Now try this

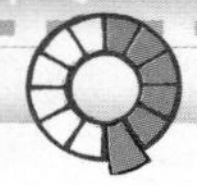

Give **two** reasons why dialysis tubing is partially permeable. **(2 marks)**

Think about how urea passes into the dialysis fluid without useful parts of the blood being lost.

Reproductive hormones

The menstrual cycle

Between puberty and about the age of 50, women have a **menstrual cycle** that repeats about every 28 days. During the cycle, changes take place in the ovaries and the uterus.

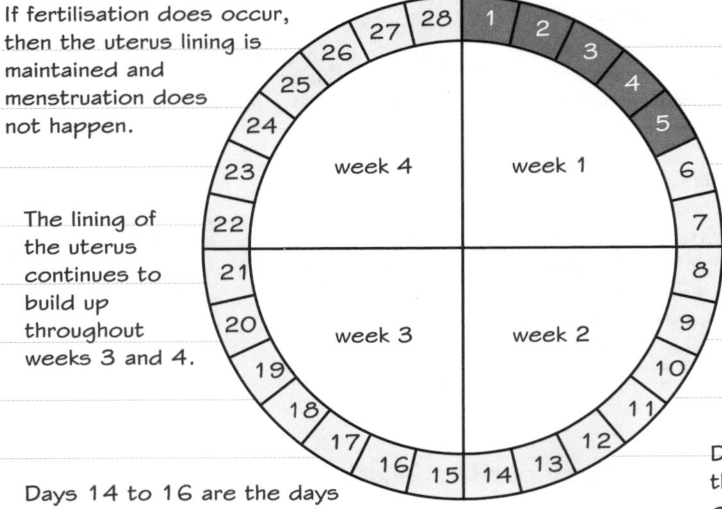

FSH, LH, oestrogen and **progesterone** are four of the hormones that control the menstrual cycle.

Hormones in the cycle

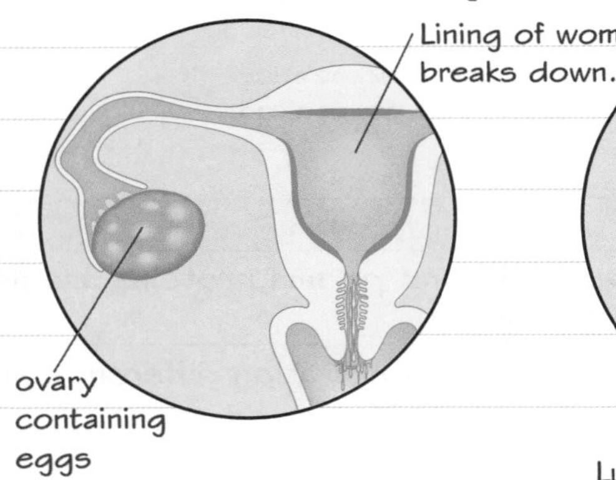

Follicle-stimulating hormone (FSH)
- causes eggs in ovaries to mature

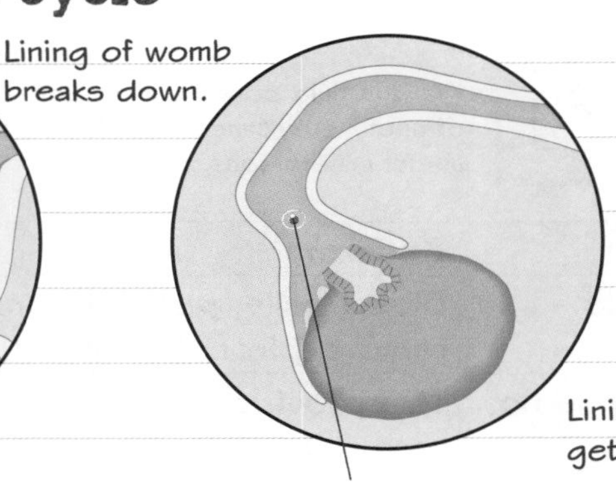

Luteinising hormone (LH)
- causes an ovary to release an egg

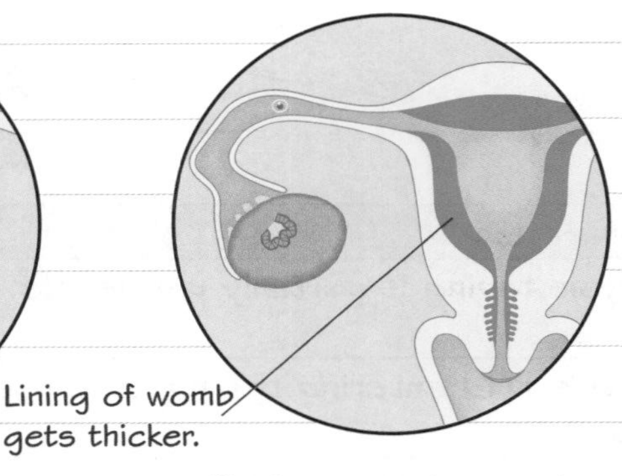

Oestrogen and progesterone
- involved in maintaining the lining of the uterus

Worked example

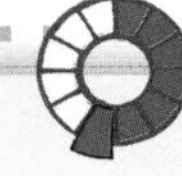

Name the main male reproductive hormone, and describe its effects at puberty. **(4 marks)**

The hormone is testosterone. It stimulates sperm production. At puberty, it triggers changes such as growth of the testes and penis, a breaking voice, and facial hair growth.

Oestrogen is the main female reproductive hormone involved in puberty. Changes include breast development and the start of the menstrual cycle.

Both boys and girls have a growth spurt, and develop underarm and pubic hair.

Now try this

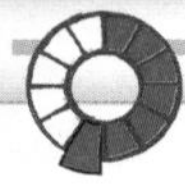

Compare the effects of FSH and LH on an egg in the ovary. **(2 marks)**

Contraception

Fertility can be controlled by hormonal and non-hormonal methods of contraception.

Oral contraceptives

Oral contraceptives may contain the hormones **oestrogen** and **progesterone**.

These hormones inhibit the release of **FSH**. This means that:

- eggs do not mature, so
- no eggs are ready for release from an ovary.

Use of oral contraceptives carries a slight risk of side effects including:

- raised blood pressure
- blood clots (thrombosis)
- breast cancer.

Worked example

Early oral contraceptives used much higher doses of oestrogen than modern ones. Some modern oral contraceptives contain only progesterone. Explain why the amounts and type of hormone were changed. **(2 marks)**

Large amounts of oestrogen caused side effects in women. The newer pills are still effective but cause fewer side effects.

Progesterone injections, implants or skin patches may allow slow release over months or years.

Surgical methods

Women are sterilised by blocking or sealing the oviducts that lead from the **ovaries** to the **uterus**. This stops eggs reaching sperm from a man and being fertilised. The **menstrual cycle** continues in women having this surgery, and they still have periods.

Men are sterilised by a **vasectomy** (see diagram). The sperm ducts are cut, stopping sperm getting into the semen.

Women and men must use other forms of contraception for a few months after surgery. The surgery is difficult to reverse.

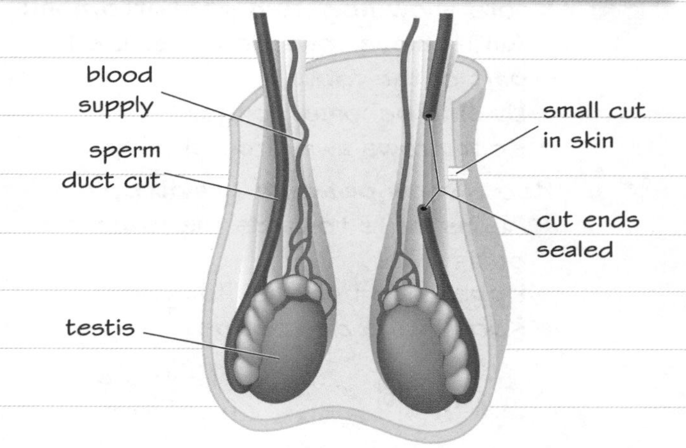

Worked example

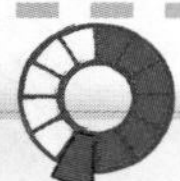

Describe **two** barrier methods of contraception. **(4 marks)**

These rely on stopping sperm reaching an egg. A condom is a thin sheath made from latex. When placed over the penis, it collects semen. A diaphragm is made of thin latex. When it is placed over the woman's cervix, it stops sperm entering the uterus.

Both methods fail if they are used incorrectly or the latex is damaged. Their effectiveness increases when used with a **spermicide**, a substance that kills or disables sperm.

Other methods

An **intrauterine device**, IUD, is a small T-shaped plastic and copper object. It is inserted into the uterus by a doctor or nurse. An IUD:

- stops sperm and eggs from surviving in the uterus or oviducts
- may also prevent a fertilised egg from implanting in the uterus.

Normal fertility returns when a doctor or nurse removes it again.

Some couples practise **natural family planning**. They abstain from intercourse at the time in the menstrual cycle when an egg may be in the oviduct.

Now try this

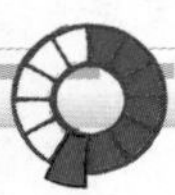

Describe **two** ways in which a woman can temporarily control her fertility. **(4 marks)**

Plant hormones

Plants produce hormones to control and coordinate their growth and development.

Tropisms

Plants respond to changes in light, moisture and gravity by changing how they grow:

- Growth in response to light is called **phototropism.**
- Growth in response to gravity is called **gravitropism** or **geotropism**.

Growth towards a stimulus is a **positive tropism**, and growth away from a stimulus is a **negative tropism**. Plant hormones called **auxins** control these tropisms.

Shoots grow:
- towards light
- against gravity.

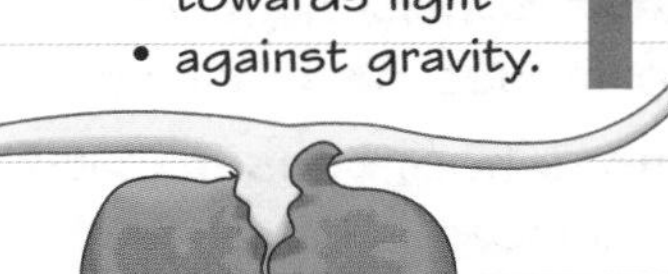

Roots grow:
- away from the light
- in the direction of gravity.

Tropisms in roots

Auxins **inhibit** cell elongation in roots.

1. Roots show **negative phototropism**:
   - Auxins move towards the shaded part of the root.
   - Lit side becomes longer.
   - Root grows away from the light.

2. Roots show **positive gravitropism**:
   - Auxins move towards the lower part of the root.
   - Upper side becomes longer.
   - Root grows downwards.

Roots and shoots grow straight if the auxin levels are equal on both sides.

Tropisms in shoots

Auxins **stimulate** cell elongation in shoots.

1. Shoots show **positive phototropism**:
   - Auxins move towards the shaded part of the shoot.
   - Shaded side becomes longer.
   - Shoot grows towards the light.

2. Shoots show **negative gravitropism**:
   - Auxins move towards the lower part of the shoot.
   - Lower side becomes longer.
   - Shoot grows upwards.

Notice that opposite processes happen in roots and shoots.

Worked example

This shoot is growing towards the light. Explain what is happening to auxin at point A, and to the cells at point B. **(3 marks)**

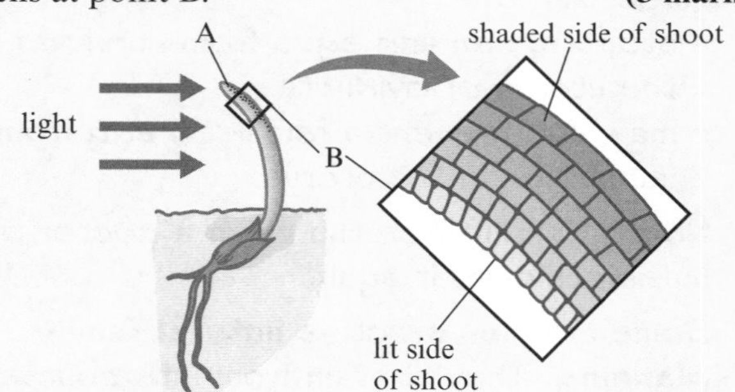

At point A, light is causing the auxin to move to the shaded side of the shoot.

At point B, cells on the shaded side contain more auxin than cells on the lit side. So the cells of the shaded side grow longer and the shoot grows towards the light.

Light is needed for **photosynthesis**. This response enables the shoot to grow towards the light, so that photosynthesis can happen at a greater rate.

Now try this

1 Describe the difference in the response of root cells and shoot cells to auxins. **(2 marks)**

2 Describe an advantage of positive phototropism to a plant. **(2 marks)**

Investigating plant responses

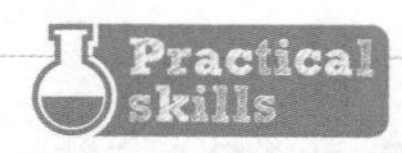

You can investigate the effect of **light** or **gravity** on the growth of newly germinated seedlings.

Core practical

Investigating the effect of light intensity on the growth of seedlings

Aim

to investigate the effect of light on the height of seedlings

Apparatus

- cress seeds
- Petri dishes
- cotton wool
- ruler, marked in 1 mm scale divisions
- water
- bright windowsill and dark cupboard

White mustard seeds could be used instead.

Dim light conditions could be achieved by placing a Petri dish on the windowsill or in the cupboard on alternate days.

Method

1 Spread some cotton wool on the bottom of three Petri dishes, and wet it with water.
2 Add about 10 seeds to each dish, then put the dishes in a warm place to allow the seeds to germinate.
3 Once the seeds have germinated, remove any excess seedlings so each dish contains the same number of seedlings.
4 Place one dish on the windowsill, one in a cupboard and one in a dim place.
5 Every day, measure the height of each seedling. Continue for at least one week.

You may need to add more water from time to time to stop the cotton wool drying out.

Rest the end of the ruler on the cotton wool and hold it gently against the seedling.

Results

Record your results in three suitable tables, showing the height of each seedling each day.

Analysis

1 Calculate the daily mean heights of the seedlings in each Petri dish.
2 Plot a graph with:
   - mean height on the vertical axis
   - day on the horizontal axis.

   Plot the results for all three Petri dishes on the same axes, and label the conditions.
3 Describe what the results show about the effect of light on the height of seedlings.

The effects of gravity

You could also investigate the effects of gravity on growth:

- Roll up a paper towel, place it in a beaker, and dampen it with water.
- Place a germinated bean seed between the paper and the glass.
- Allow it to grow so the root and shoot are visible, then lie the beaker on its side.

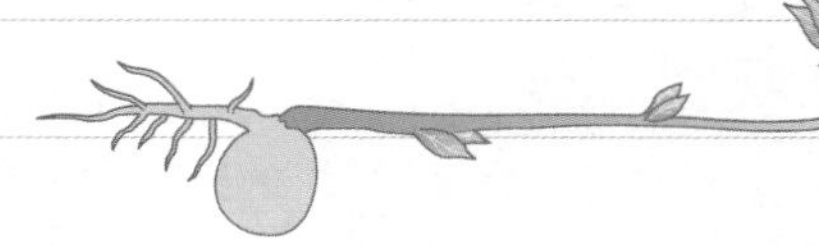

Observe and record (with labelled biological drawings) what happens to the root and shoot over the next few days.

Now try this

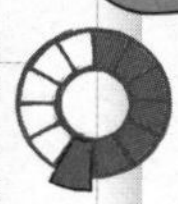

1 Describe why each Petri dish must contain several seedlings. **(2 marks)**

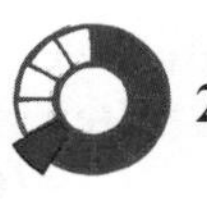

2 The scale on most rulers begins a few millimetres from the end. Explain the effect of this on the accuracy of the height measurements. **(3 marks)**

Extended response – Homeostasis and response

There will be at least one 6-mark question on your exam paper. For these questions, you will need to think scientifically and structure your answer logically, showing how the points you make are related to each other. You can revise the topic for this question, which is about **neurones**, on page 51.

Worked example

Motor neurones are one type of cell found in the nervous system. The diagram shows the structure of a motor neurone.

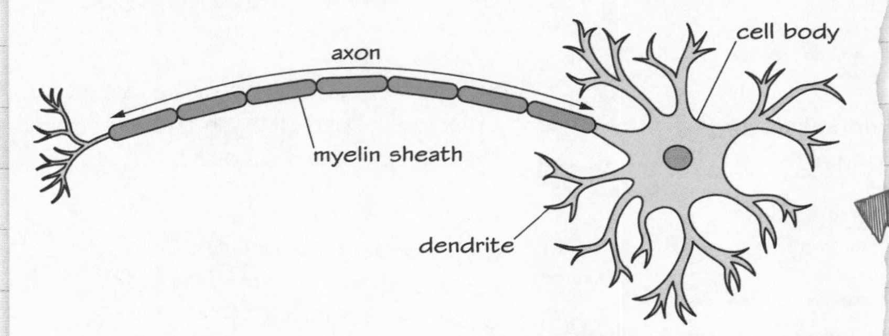

Explain how the structure of a motor neurone is related to its function. **(6 marks)**

Neurones carry nervous impulses along their axons. Motor neurones carry impulses from relay neurones to effectors such as muscles or glands. They have long axons to carry impulses long distances through the body. The myelin sheath insulates the axon, making the impulses travel faster.

Motor neurones have many tiny dendrites at the end of the axon. These receive impulses from other neurones in the nervous system, connecting neurones together.

The gap between one neurone and the next is called a synapse. Chemicals released into the synapse allow the signal to pass to the next neurone.

Command word: Explain

When you are asked to **explain** something, it is not enough just to state or describe it.

Your answer **must** contain some reasoning or justification of the points you make.

Diagrams are shown in questions to give you information that will be useful in your answers. Make sure you study them carefully.

The question does not ask for the function of motor neurones, but the answer includes the function. This is important for explaining how the structure and function are related.

The answer mentions nervous impulses, but they could be called signals or messages instead.

This part of the answer includes a description of the synapse. It is not part of the cell itself, but an understanding of the synapse is needed to fully explain the function of a motor neurone.

You need to show comprehensive knowledge and understanding, using relevant scientific ideas, to support your explanations. In this answer, the structure of the three main parts of the motor neurone are described and explained (the axon, myelin sheath and dendrites).

Now try this

When a bright light shines into the eye, muscles in the iris contract. The size of the pupil decreases so that less light can enter the eye. This is an example of a reflex. Describe how impulses travel through the reflex arc involved, starting with receptor cells in the eye. **(6 marks)**

Think about the route, including the three types of neurones involved and the effector. You can revise this topic on page 52.

Meiosis

Cells in reproductive organs divide by **meiosis** to form **gametes**.

Meiosis outlined

When a cell divides by meiosis, the four cells produced are all genetically different.

The parent cell is a diploid cell. So it has two sets of chromosomes.

The parent cell divides in two and then in two again. Four daughter cells are produced.

Each chromosome is copied.

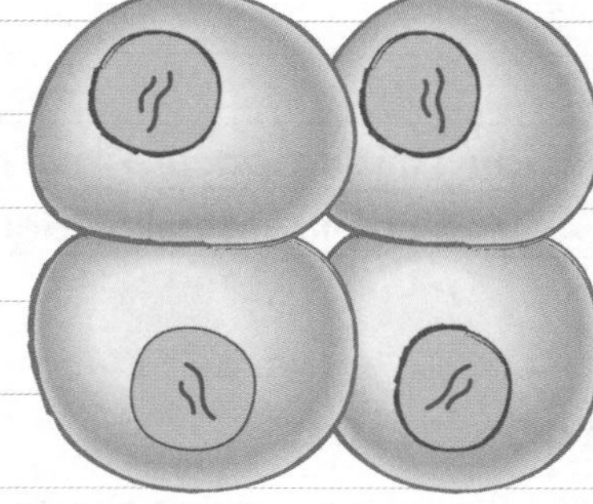

Each daughter cell gets a copy of one chromosome from each pair.

parent cell

The cells produced by division are always called 'daughter cells' even if they will eventually turn into sperm cells.

Each daughter cell has only one set of chromosomes. So these are haploid cells.

Fertilisation

At **fertilisation**, a male gamete joins with a female gamete to form a new cell.

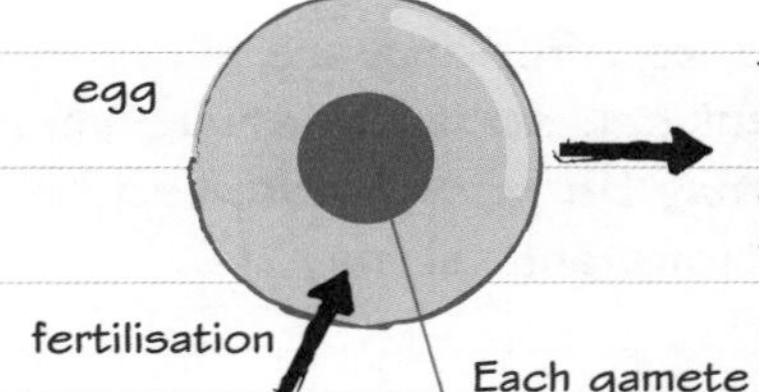

The new cell is called a zygote. It has two sets of chromosomes, like all body cells (human body cells have 46 chromosomes).

Zygote divides repeatedly by mitosis, forming an embryo. The number of cells increases and the cells differentiate as the embryo develops.

Each gamete has one set of chromosomes (human gametes have 23 chromosomes).

Sexual reproduction produces **variation** in the offspring, as the fertilised cell contains one set of chromosomes from each parent. So one of each pair of alleles in each body cell comes from each parent. (An allele is one form of a gene.)

Notice that:

- meiosis produces gametes with half the normal number of chromosomes
- fertilisation restores the normal number of chromosomes.

Worked example

Compare mitosis and meiosis. **(4 marks)**

You can revise mitosis on page 9.

Meiosis produces four daughter cells, but mitosis only produces two.

Meiosis produces genetically different daughter cells, but in mitosis the daughter cells are genetically identical to each other and to the parent cell.

Meiosis occurs in reproductive organs and produces haploid gamete cells. Mitosis occurs in body cells and produces diploid cells.

Now try this

Compare haploid cells and diploid cells. **(4 marks)**

Sexual and asexual reproduction

New individuals can form by sexual reproduction or by asexual reproduction.

Features of reproduction

	Sexual	Asexual
Number of parents	two	one
Meiosis occurs?	✓	✗
Gametes fuse?	✓	✗
Mitosis occurs?	✓	✓
Genetic information	mixing occurs	no mixing
Offspring	not identical	identical

Gametes form by meiosis in sexual reproduction:
- Male gametes are **sperm** cells in animals and **pollen** cells in flowering plants.
- Female gametes are **egg** cells in animals and flowering plants.

Mitosis forms **clones**, genetically identical offspring, in asexual reproduction. It increases the number of cells in both types of reproduction.

Sexual reproduction

Advantages of sexual reproduction include:
- **It** produces **variation** in offspring.
- If the environment changes, variation gives the species a survival advantage by natural selection (see page 74).
- Humans can increase food production by using selective breeding to speed up natural selection (see page 75).

Disadvantages include:
- Individuals must find a mate, which takes time and energy.

Asexual reproduction

Advantages of asexual reproduction include:
- Individuals do not need a mate, which is efficient in terms of time and energy.
- It is faster than sexual reproduction, so many offspring can be produced quickly when the environment is favourable.

Disadvantages include:
- The parent and its offspring are all genetically identical, so if the environment changes they may be poorly adapted to the new conditions and all may die.

Worked example

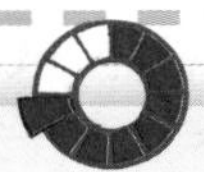

Many plants produce seeds by sexual reproduction, but also reproduce asexually. For example, strawberry plants produce runners with small plants along them. Daffodils produce underground bulbs which divide to form new bulbs. Suggest the advantages of reproducing like this. **(4 marks)**

A strawberry plant that is adapted to its environment can quickly spread by asexual reproduction. Daffodils can form new plants, even if some are damaged by frost and fail to be pollinated so they can form seeds. The use of sexual reproduction introduces variation in the plant population, making it more likely to survive changes in conditions.

Plants usually have adaptations that cause their seeds to fall far from the parent plant. This helps to spread the species. It also reduces competition between the parent and its offspring for light, space, water and other resources.

Fungi and *Plasmodium*

Many fungi reproduce asexually by producing spores in their familiar mushrooms and toadstools. They also reproduce sexually to give variation in the population.

Plasmodium, the parasite that causes malaria, reproduces asexually in its human host but sexually in the mosquito vector.

Now try this

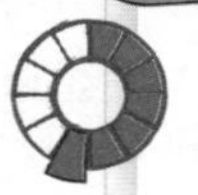

1 Describe **two** ways in which plants naturally produce clones. **(3 marks)**

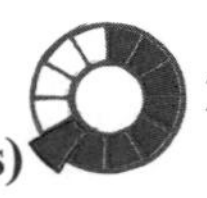

2 Explain how reproduction can introduce variation into a population. **(4 marks)**

DNA and the genome

DNA is genetic material. An organism's **genome** is its entire genetic material.

DNA structure

The nucleus contains chromosomes.

Most cells have a nucleus.

eukaryotic cell

chromosome

DNA is the genetic material. It is tightly coiled in a chromosome.

A **gene** is a small section of DNA on a chromosome. Each gene codes for a particular sequence of amino acids to make a specific protein.

DNA

DNA is a polymer made up of two strands, forming a double helix structure.

Worked example

The diagram shows a short section of DNA.

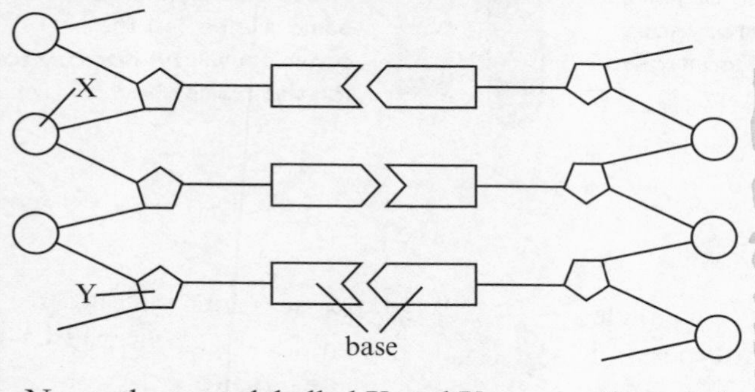

Name the parts labelled **X** and **Y**. **(2 marks)**

X is a phosphate group and Y is a sugar.

DNA is made up of four different repeating **nucleotide** units. A nucleotide consists of a phosphate group and a base attached to a sugar. Each DNA strand has alternating phosphate and sugar parts.

There are four different **bases** (C, G, T, A). The code for an amino acid is a sequence of three bases. The order of bases controls the order that amino acids join to form a protein.

The human genome

The whole human genome has now been studied. This allows scientists and doctors to:

1. search for genes linked to different types of disease. This could help to identify people at increased risk of developing diseases such as cardiovascular disease or some cancers.
2. understand and treat inherited disorders, such as cystic fibrosis, in new ways. You can revise inherited disorders on page 73.

Human migration

Some parts of human DNA remain unchanged from one generation to the next. Scientists analyse these 'markers' in populations around the world. They use these data to produce maps showing how humans migrated in the past.

Mitochondria have their own DNA. We inherit our mother's mitochondria, so the maternal line can be analysed. Men inherit their Y chromosome from their father, so the paternal line can be analysed.

Now try this

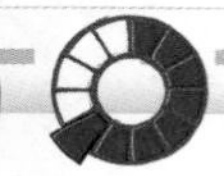

Describe the structure of DNA. **(5 marks)**

Genetic terms

You need to be able to explain all the **genetic terms** shown in bold on this page.

Inside a cell

A **gamete** is a sex cell formed by meiosis in a reproductive organ.

At fertilisation:

- sperm cells and egg cells fuse (join together) in animals
- pollen cells and egg cells fuse in flowering plants.

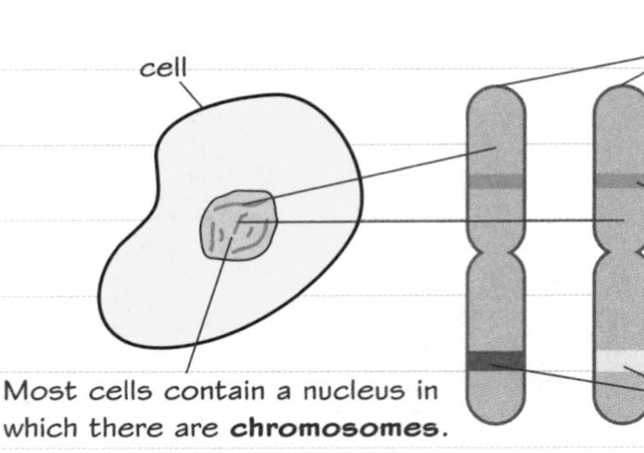

There are two copies of each **chromosome** in body cells – each copy has the same genes in the same order along its length (except chromosomes that determine sex).

A **gene** is a small section of DNA on a chromosome. Each gene codes for a particular sequence of amino acids to make a specific protein.

A gene may come in different forms, called **alleles**, that produce different variations of the characteristic, e.g. different eye colours.

Alleles

The **genotype** is the **alleles** of a particular gene present in an individual. The **phenotype** is an individual's observed characteristics or traits, produced by these alleles working at a molecular level.

Most characteristics are a result of multiple **genes** interacting, rather than the result of a single gene.

Here there are different **alleles** of the same gene, so the organism will be **heterozygous** for the trait controlled by the gene.

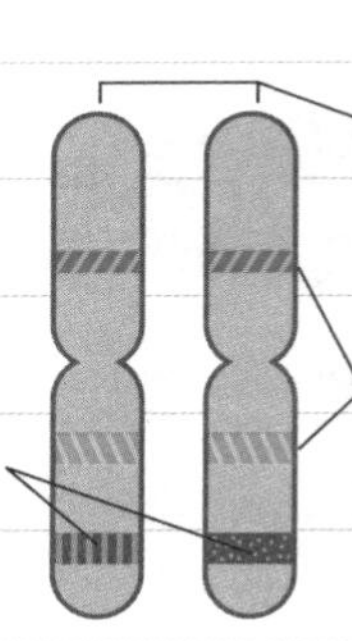

Chromosomes of the same pair have the same genes in the same order.

These two genes have the same alleles, so the organism will be **homozygous** for the traits they control.

Worked example

Fur colour in mice is controlled by a single gene. This has two alleles, shown as B and b. The table shows the three possible genotypes and phenotypes for this trait.

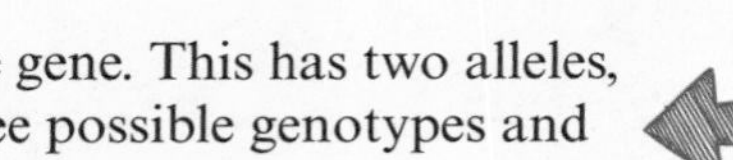

Red–green colour blindness in humans is also controlled by a single **gene**.

Genotype	Phenotype
BB	black fur
Bb	black fur
bb	brown fur

(a) Explain why the allele for black fur is described as a **dominant** allele. **(2 marks)**

Mice have black fur whether they have two copies of this allele, BB, or just one, Bb.

A **dominant allele** is always expressed, even if only one copy is present:

- BB mice are homozygous for the B allele.
- Bb mice are **heterozygous**.

(b) Explain why the allele for brown fur is described as a **recessive** allele. **(2 marks)**

Mice only have brown fur when they have two copies of this allele, bb. They have black fur instead if they only have one copy, Bb.

A **recessive allele** is only expressed if two copies are present (so no dominant allele):

- bb mice are **homozygous** for the b allele.

Now try this

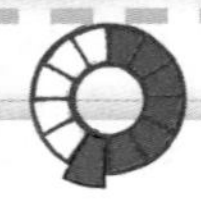

Flower colour in pea plants is controlled by two alleles. R is the dominant allele and r is the recessive allele. Describe the difference between the two types of allele. **(2 marks)**

Genetic crosses

The results of **genetic crosses** are predicted using genetic diagrams and Punnett square diagrams.

Genetic diagrams

In the **genetic diagram** on the right, both parent plants are heterozygous for flower colour. It shows that the possible combinations of alleles for this Rr × Rr genetic cross are:

- RR, purple flowers
- Rr, purple flowers (two ways to get this)
- rr, white flowers.

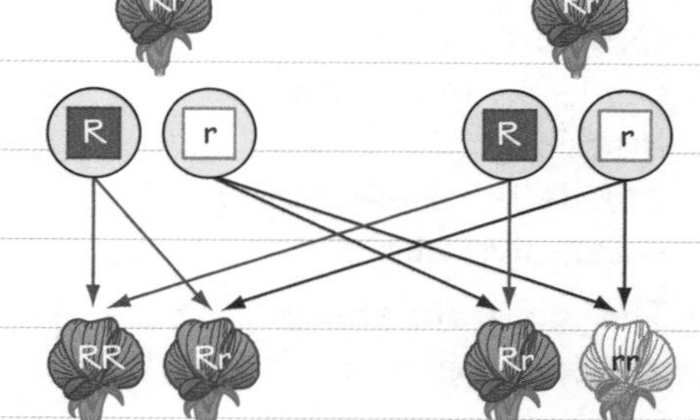

Probability

Notice in the two diagrams on the right that:

- ¼ combinations are RR (purple flowers)
- ¼ combinations are rr (white flowers).

There are two combinations that give Rr, so:

- ²⁄₄ combinations are Rr (purple flowers).

The **probability** that these parents will produce offspring with purple flowers is:

$(1/4 + 2/4) = 3/4$

You could also write this as 0.75, 75%, 3 in 4 or 3:1. The actual outcome may differ, particularly if few offspring are produced.

Punnett square diagrams

You can also show the outcomes of the Rr × Rr genetic cross above using a **Punnett square diagram**.

		Parent gametes	
		R	r
Parent gametes	R	RR	Rr
	r	Rr	rr

You should be able to complete or draw diagrams like these.

Worked example

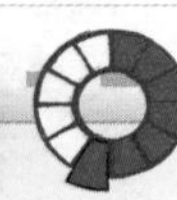

The allele for green seed pods, G, is dominant to the allele for yellow seed pods, g.

(a) Complete the Punnett square diagram to show the possible outcomes of a Gg × gg genetic cross. **(1 mark)**

Gametes	G	g
g	Gg	gg
g	Gg	gg

It may help the answer to part (b) if the seed pod colour is written underneath each outcome.

(b) Identify the offspring that will produce yellow seed pods, and explain your answer. **(2 marks)**

The gg offspring will have yellow seed pods. This is because the allele for yellow, g, is recessive. It will only be expressed if an individual has two copies.

The probability of having an offspring that produces yellow seed pods is ²⁄₄. You could also write this as 0.5, 50%, 1 in 2 or 1:1.

Now try this

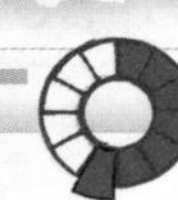

The allele for black fur in rabbits, B, is dominant. The allele for brown fur, b, is recessive.

(a) Complete the Punnett square diagram.

Gametes	B	B
b		
b		

(1 mark)

(b) State the probability that this cross will produce baby rabbits with brown fur, and explain your answer. **(3 marks)**

Family trees

Family trees can be analysed to study the inheritance of alleles.

Analysing a family tree

Cystic fibrosis, CF, is an inherited disorder (see page 73). It is caused by a recessive allele.

The diagram shows an example of a family tree in which some members have cystic fibrosis.

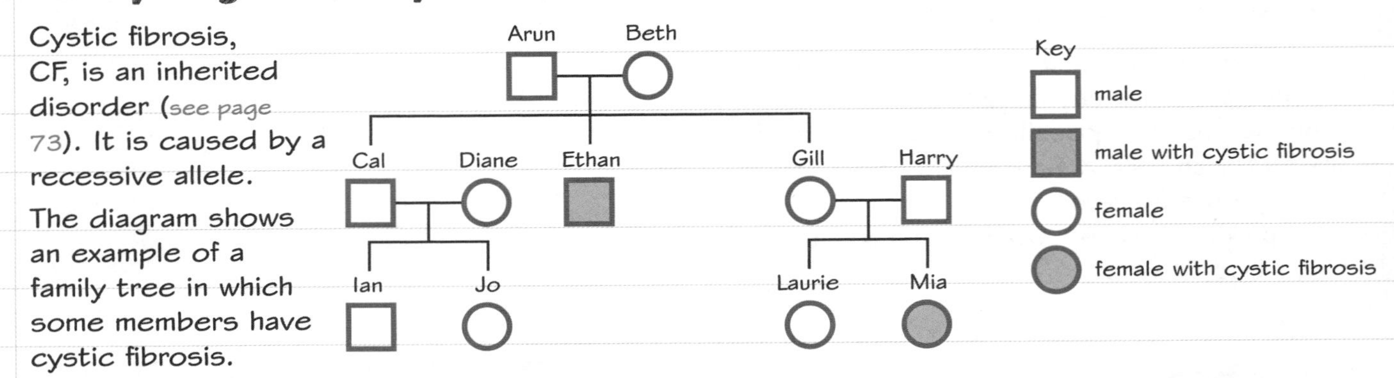

Ethan, one of Arun and Beth's three children, has cystic fibrosis:

- The allele for cystic fibrosis is recessive, so Ethan must have inherited one copy from each parent.
- Arun and Beth must be heterozygous for the cystic fibrosis allele.

Mia, one of Gill and Harry's two daughters, has cystic fibrosis:

- The allele for cystic fibrosis is recessive, so Mia must have inherited one copy from each parent.
- Gill and Harry must be heterozygous for the cystic fibrosis allele.

Worked example

The ability to taste phenylthiocarbamide (PTC) is a dominant condition. The diagram shows the inheritance of PTC tasting in a family. Describe the evidence that PTC tasting is controlled by a dominant allele. **(2 marks)**

male female
1 2 3 4
5 6 7 8 9 10 11
12 13 14 15 16
non-PTC tasters PTC tasters

Individuals 1 and 2 are PTC tasters, but they have two children (individuals 6 and 8) who are non-tasters. Therefore, PTC tasting must be controlled by a dominant allele and non-tasting by a recessive allele.

Trees and probability

You can use a **Punnett square diagram** (page 71) to calculate the probability that Gill and Harry will have a child with cystic fibrosis.

Gametes	F	f
F	FF	Ff
f	Ff	ff

A child with cystic fibrosis will be ff, as the allele is recessive so two copies must be present.

There is one ff combination from the four possible combinations, so the **probability** of having a child with cystic fibrosis is:

¼ (0.25, 25%, 1 in 4 or 1:3).

In this family, the actual outcome is ½ (0.5, 50%, 1 in 2 or 1:1).

Now try this

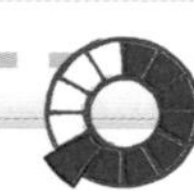

The question is about the family tree in the worked example.

(a) Give the genotypes for individuals 9 and 10, and explain your answer. **(4 marks)**

(b) Explain why individuals 5 and 6 can only produce children who are non-PTC tasters. **(4 marks)**

Use T for the dominant allele and t for the recessive allele.

Think about the type of allele that controls the ability to taste PTC.

Inheritance

Some disorders and the sex of an individual are inherited.

Polydactyly

Polydactyly is an **inherited disorder** in which people have extra fingers or toes. It is caused by a dominant allele. This means that:

- people have polydactyly if they inherit one **or** two copies of the allele.

The Punnett square diagrams show the two situations where one parent has polydactyly.

Polydactyly does not usually harm health. Other dominant genetic disorders may lead to ill-health, such as Huntington's disease.

If both parents are homozygous (one with polydactyly and one without):

- all the children will be heterozygous so will have polydactyly.

Gametes	P	P
p	Pp	Pp
p	Pp	Pp

If one parent is heterozygous (one with polydactyly and one without):

- the probability of having a child with polydactyly is ½ (0.5, 50%, 1 in 2 or 1:1).

Gametes	P	p
p	Pp	pp
p	Pp	pp

Cystic fibrosis

Cystic fibrosis is an inherited disorder of the cell membranes. It is caused by a recessive allele. This means that:

- People have cystic fibrosis only if they inherit two copies of the allele.
- People who are heterozygous for the trait are carriers – they can pass the allele to their children but do not have the disorder themselves.

If both parents are heterozygous (carriers):

- the probability of having a child with cystic fibrosis is ¼ (0.25, 25%, 1 in 4 or 1:3).

Gametes	F	f
F	FF	Ff
f	Ff	ff

carrier (like the parents) → Ff

child with cystic fibrosis → ff

Worked example

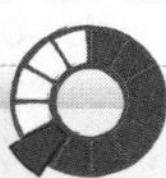

Draw a Punnett square diagram to show how sex in humans is inherited. Identify the sex of each offspring in your diagram. **(3 marks)**

Father \ Mother	X	X
X	XX female	XX female
Y	XY male	XY male

Ordinary human body cells have 23 pairs of chromosomes. One of these pairs carries the genes that determine sex:

- females – same sex chromosomes, XX
- males – different sex chromosomes, XY.

The probability of a couple having a girl is ½ or 50%. This is the same as the probability of having a boy.

Now try this

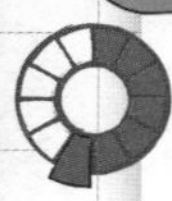

1 Explain why all human egg cells contain the same sex chromosome. **(2 marks)**

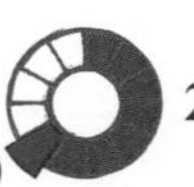

2 Describe **two** differences between the inheritance of polydactyly and the inheritance of cystic fibrosis. **(2 marks)**

Variation and evolution

Evolution is a change in the inherited characteristics of a population over time due to natural selection. It may result in the formation of a new species.

Causes of variation

A **population** is all the individuals of a species living in the same place at the same time.

The differences in the characteristics of the individuals in a population are called **variation**.

Most variation is caused by a combination of genes and environment.

Variation differences between individuals in a population

Genetic causes differences due to the genes inherited by the organism, e.g. eye colour

combination of genes and the environment, e.g. body mass. height

Environmental causes differences in the conditions in which the organism developed, e.g. scars, language

Evolution

The theory of evolution by **natural selection** states that all species of living things have evolved from simple life forms, which first developed over three billion years ago.

Individuals in a population show variation in their inherited characteristics.

↓

Some individuals have phenotypes that make them better suited to their environment.

↓

These individuals are more likely to survive and reproduce, so pass their alleles to their offspring.

↓

These offspring are more likely to have phenotypes that make them better suited to the environment. They are more likely to survive and reproduce.

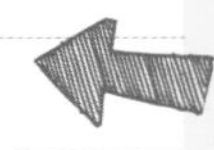

Two populations of one species may become too different to breed with each other any more to produce fertile offspring. They will have formed two species.

Variation and mutation

There is usually extensive **genetic variation** within a population of a species.

All genetic **variants** arise from **mutations**. Of these mutations:

- most have no effect on the **phenotype**
- some influence the phenotype
- very few determine the phenotype.

Mutations occur continuously, but very rarely a mutation may lead to a new phenotype. If this is suited to an environmental change, it can lead to a relatively rapid change in a species.

Worked example

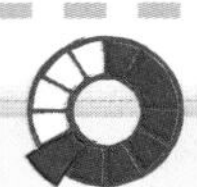

This is an example of where you may need to apply your knowledge and understanding to an unfamiliar example or situation.

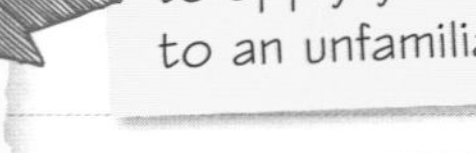

Poachers hunt African elephants illegally for their tusks. A mutation may cause elephants to be born tuskless. In Uganda in 1930, only 1% of elephants were born tuskless. By 2010, 15% of female elephants and 9% of male elephants were born without tusks. Explain this change using the theory of evolution. **(3 marks)**

Elephants with tusks are more likely to be poached and killed, so do not survive to pass on their alleles. Tuskless elephants are more likely to survive to pass on the allele for no tusks, so the allele for no tusks increases in frequency in the population.

Now try this

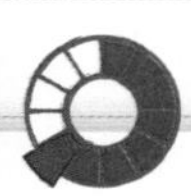

1 Identical twins are produced when a fertilised egg cell divides in two. Explain why identical twins share many, but not all, of their characteristics. **(3 marks)**

2 Head lice can be killed using shampoos containing insecticides. Some strains of head lice are resistant to some insecticides. Explain the advantage of this to these head lice. **(3 marks)**

Selective breeding

Humans have been using **selective breeding** of plants and animals for thousands of years.

The process

Selective breeding is also called **artificial selection**. Humans use this process to breed plants and animals that have particular genetic characteristics. The process of selective breeding takes place over many generations, not just one.

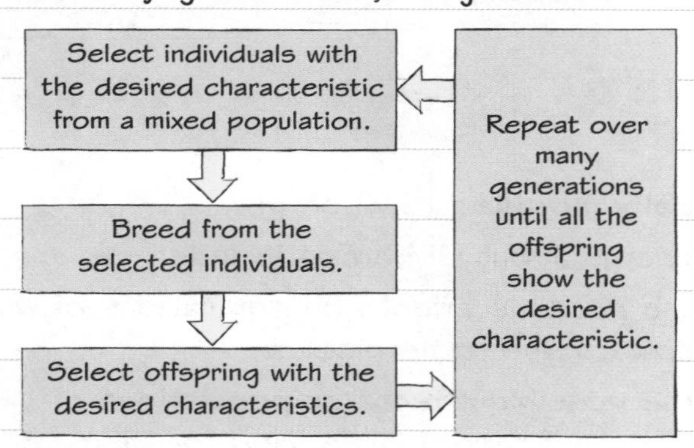

Worked example

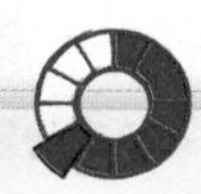

Describe how a farmer would produce a herd of cattle that has a high milk yield. **(4 marks)**

The farmer would choose a cow that produced a lot of milk. This cow would then be bred with a bull whose mother also had a high milk yield. Next, the farmer would choose the female offspring that have a high milk yield, and breed them with a bull whose mother also had a high milk yield. The farmer would do this for many generations.

Characteristics can be chosen for usefulness or appearance, including:
- animals that produce more milk or meat
- food crops that resist plant diseases
- domestic dogs that have a gentle nature
- plants with large or unusual flowers.

Worked example

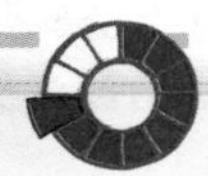

The graph shows the mean milk yield per cow from 1957 to 2007. The change is mainly the result of selective breeding. Over the same period, there was an increase in the percentage of cows with mastitis (sore udders) and leg problems. This was reported in a farming newspaper in an article with the title 'Selective breeding has improved dairy farming'.

Evaluate the statement in the title. **(5 marks)**

Milk yield per cow/kg: 4000, 5000, 6000, 7000, 8000, 9000
Year: 1957, 1967, 1977, 1987, 1997, 2007

The graph shows that the milk yield per cow has more than doubled between 1957 and 2007. This is useful because farmers can produce more milk from fewer cows, reducing costs.

This is an advantage to the farmer.

However, more cows are having problems with their legs and udders. This may be because they are producing so much milk, so this is an animal welfare issue. It will also cost the farmer more in veterinary bills.

This describes an **ethical issue** and a disadvantage to the farmer.

Overall, I think that selective breeding has improved dairy farming, but the animal welfare issues it has caused must be sorted out.

Your conclusion should relate to information given, and benefits and risks identified.

Inbreeding

Selective breeding, especially if it involves close relatives, can lead to **inbreeding**. Some breeds may be particularly likely to suffer from disease or inherited defects.

Now try this

Describe how a breed of sheep that grow a lot of wool could be produced. **(4 marks)**

Genetic engineering

Genetic engineering is a process in which an organism's **genome** is modified to give a desired characteristic by introducing a gene from another organism.

The process

The flow chart outlines the process of genetic engineering.

Genes can be transferred from any kind of organism to any other kind of organism, e.g. bacteria, humans, other animals, plants.

The gene for a characteristic is 'cut out' of a chromosome using enzymes. → The gene is transferred to a cell of another organism, and inserted into a chromosome. → The cell of this organism now produces the characteristic from the gene.

GM crops

Genetically engineered crop plants are called **genetically modified** (GM) crops.

GM crops include plants that are:

- resistant to plant diseases
- resistant to attack by insect pests
- resistant to herbicides, so that fields can be sprayed to kill weeds, but not the crop plants.

These characteristics help to produce increased **yields**, including bigger and better fruits.

Concerns about GM crops include:

- the possible effects on populations of wild flowers due to herbicides
- the possible effects on populations of insects if there is less food for them
- whether the possible effects on human health of eating GM crops has been researched enough.

Worked example

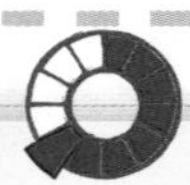

Patients with diabetes may be treated with human insulin. Bacteria have been genetically engineered to carry the gene for human insulin. These bacteria are grown in large numbers to produce relatively large amounts of human insulin.

(a) Explain why the bacteria had to be modified to make human insulin. **(2 marks)**

Bacteria do not naturally have the gene to produce human insulin. Genetic engineering must be used to introduce the gene into their genome.

(b) Describe how the genetically engineered bacteria were produced. **(2 marks)**

The gene for making human insulin was cut out of a human chromosome. The gene was then inserted into bacterial cells so it would become part of their genome. The modified bacteria would then be able to make insulin.

Engineering sheep

Sheep have been genetically engineered to produce a human protein in their milk. Some people have an inherited disorder that stops them making this protein, causing lung and liver problems. The protein in sheep's milk can be used to treat them.

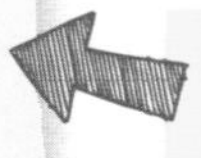

Scientists are researching ways to genetically modify cells in people with inherited disorders. They hope that this treatment would overcome the symptoms of these disorders. However, some people have objections to genetic engineering, including ethical concerns.

Now try this

1 Give the meaning of **genetic engineering**. **(3 marks)**

2 Describe how a GM crop with herbicide resistance could be developed. **(3 marks)**

Cloning

Plants and animals can be **cloned** to produce large numbers of genetically identical individuals.

Cloning plants

The diagrams show the stages involved in **tissue culture**. This is important for:
- preserving rare species of plants
- producing many plants in plant nurseries.

- Gardeners can produce many identical new plants by taking **cuttings** from a parent plant.

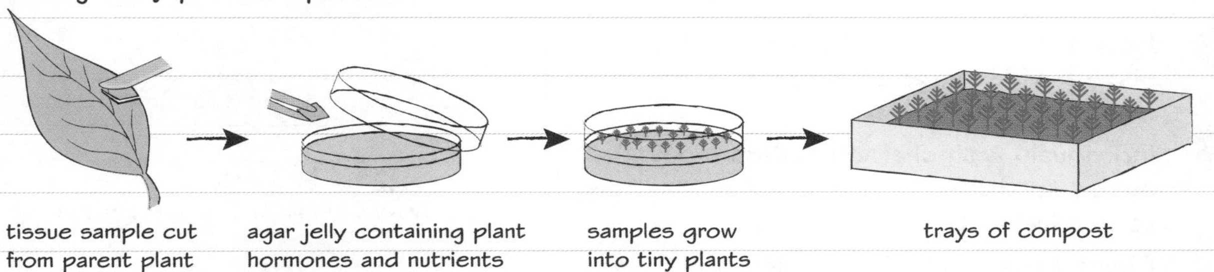

Embryo transplant

This method of cloning animals involves:
- separating cells from a developing embryo at a stage before they become specialised
- transplanting these genetically identical embryos into host mothers.

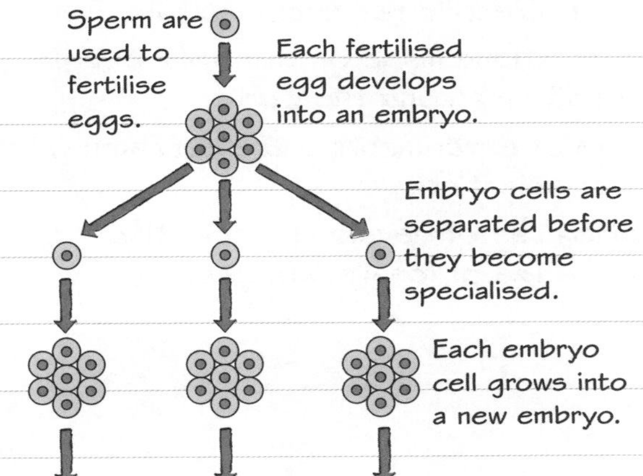

The offspring are clones of the original embryo.

Adult cell cloning

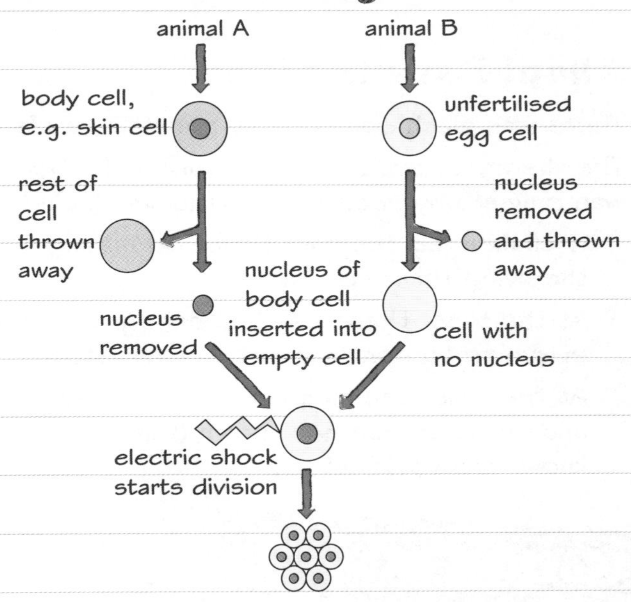

- Cell divides to form embryo.
- Embryo is inserted into womb of adult female C to develop until birth.

Worked example

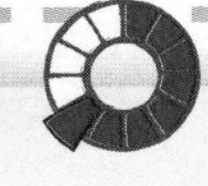

The diagram above shows how adult cell cloning works. The offspring is a clone of one of the three animals involved. Give the letter of this animal, and explain your answer. **(2 marks)**

Animal A. This is because its genes are contained in the nucleus taken from the body cell of animal A.

The offspring inherits no components from its host mother, animal C.

The offspring inherits its mitochondria from animal B. Mitochondria have small amounts of their own DNA, so it is not fully genetically identical to animal A.

Now try this

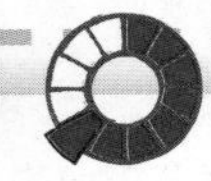

Explain which method of animal cloning is likely to be used to produce many clones. **(2 marks)**

Darwin and Lamarck

The theory of evolution by natural selection developed over time using many scientists' work.

Charles Darwin (1809–1882)

Charles Darwin proposed the theory of **evolution** by **natural selection**. He published his ideas in his book *On the Origin of Species* in 1859. The main points of the theory are:

1. Individual organisms within a species show a wide range of variation for a characteristic.
2. Individuals with characteristics most suited to the environment are more likely to survive to breed successfully.
3. Characteristics that have enabled these individuals to survive are passed on to the next generation.

A timeline of events

1831–1836
The voyage of *HMS Beagle*. Darwin studies animals on the Galapagos Islands, including various finches.

1837
Darwin presents his collection of specimens to the Zoological Society.

1859
On the Origin of Species is published.

1871
The Descent of Man is published.

About Darwin's ideas

There was much controversy about Darwin's ideas, which were seen as revolutionary.

The theory of evolution by natural selection was only slowly accepted. Criticisms included:

- It challenged the idea that God made all the living things on Earth.
- At the time, there was not enough evidence to convince many scientists.
- At the time, the mechanism of inheritance and variation was not known (this was known 50 years later).

Although his theory was criticised, it was a result of scientific research, including:

- observations made on the *HMS Beagle* expedition around the world
- years of experiments and discussion with other scientists
- making use of developing scientific knowledge of fossils and geology.

Worked example

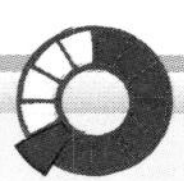

Jean-Baptiste Lamarck (1744–1829) proposed a theory of evolution by the inheritance of acquired characteristics:

- Body parts that are used become larger and stronger, while unused parts become smaller or disappear.
- Changes to an organism during its lifetime are passed to its offspring.

Use the information to suggest how Lamarck might have explained the evolution of long necks in giraffes. **(3 marks)**

A giraffe stretches upwards to reach leaves in high branches. The giraffe's neck becomes longer because it is used a lot. The giraffe's offspring inherit its long neck. Over time, giraffes develop increasingly long necks.

Alternative theories, like this one, are based mainly on the idea that changes to an organism during its lifetime can be inherited. We now know that this cannot happen in the vast majority of cases.

Now try this

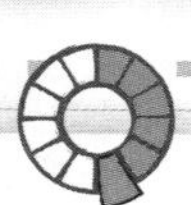

Give **three** reasons why the theory of evolution by natural selection was only gradually accepted. **(3 marks)**

Speciation

Speciation is the formation of new species.

Forming new species

Populations of the same species become separated e.g.
- they move to different islands
- a change in the environment produces a barrier to movement.

Other examples of **geographical isolation** include:
- new rivers forming
- volcanic eruptions
- mountain-building processes.

Natural selection in the different areas favours different variations of characteristics.

The original population will show variation between individuals due to genetic causes.

The characteristics of each population change over time.

Natural selection acting in the different environments produces gradual changes in each population.

Their characteristics become so different that the populations become different species.

Individuals from two different species are usually unable to produce fertile offspring.

Alfred Wallace (1823–1913)

Alfred Russel Wallace proposed the theory of evolution by natural selection:
- Like Charles Darwin, Wallace worked worldwide to gather evidence
- Wallace developed his theory independently of Darwin.

Later, in 1858, he published work jointly with Darwin. This prompted Darwin to publish his book *On the Origin of Species* a year later (you can revise Darwin's work on page 78).

Other work by Wallace

Wallace is best known for:
- his work on **warning colours** in animals such as caterpillars that are toxic to predators
- his theory of speciation.

Wallace proposed that, as two populations gradually differed, any **hybrids** would be less well adapted than their parents. This 'Wallace effect' would continue to drive speciation.

Scientists have gathered more evidence since, leading to our current scientific understanding.

Worked example

Anole lizards live on islands in the Caribbean. Each island has its own anole species. Islands that are further from the other islands have more species that are not found on other islands. DNA evidence suggests that many of the species are very closely related. Describe one way in which these species may have evolved. **(4 marks)**

A few lizards of one species moved from one island to another. The environment was different on this island, so natural selection there favoured different variations. Over time, the characteristics of the population on the new island changed enough so the lizards became a new species.

Now try this

Many species of flightless birds live on the islands of New Zealand. All these species are related to species found in nearby Australia that can still fly. Suggest how the flightless bird species may have evolved. **(4 marks)**

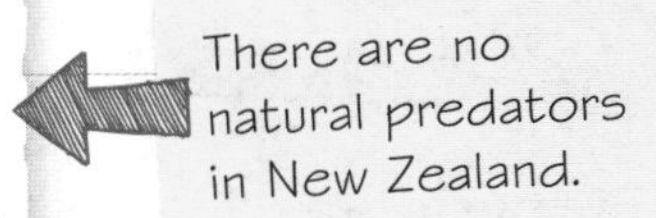

Mendel

Gregor Mendel (1822–1884) was the first person to discover the basics of **inheritance**.

Mendel's work

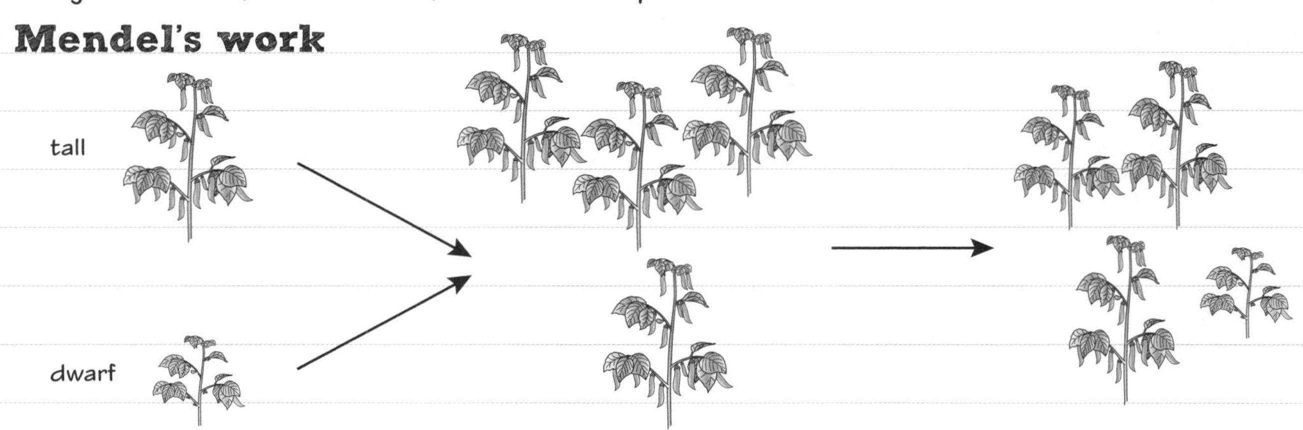

Pure-bred peas	1st generation	2nd generation
Mendel's pure-bred pea plants would always produce identical offspring when bred with the same type of pea plant.	When Mendel crossed a tall plant with a dwarf plant, all the offspring grew to be tall plants. The dwarf trait was recessive to the dominant tall trait.	Mendel then cross-bred plants from the first generation with each other. About one quarter of the offspring were dwarf. The dwarf trait had been passed down unchanged from the pure-bred parents.

- Mendel was the first person to show that, in the second generation, there were about three times more plants with the dominant characteristic than plants with the recessive characteristic.
- Few people read Mendel's work, published in 1865.
- It was only many years later that scientists linked Mendel's 'units' (inherited characteristics) with genes.

A timeline of discovery

The gene theory was developed from the work of many scientists over a long time.

Mid-19th century	Late 19th century	Early 20th century	Mid-20th century
Mendel shows that the inheritance of a characteristic is determined by 'units' that pass on unchanged.	The behaviour of chromosomes during cell division is observed.	'Units' and chromosomes are seen to behave in similar ways, so 'units' are genes located on chromosomes.	The structure of DNA is determined, and the mechanism of gene function is worked out.

Worked example

Mendel concluded from his experiments with pea plants that characteristics are caused by 'units' – separately inherited factors that were passed unchanged to descendants.
Explain Mendel's conclusion. **(2 marks)**

Mendel showed that some characteristics of parent plants, such as dwarf size, seem to disappear in the first generation but reappear in the second generation. So the factor that causes this characteristic must be kept separate, even in the tall first generation plants.

Mendel used pure-bred parents. He also:
- made the crosses by hand so he knew which parents produced which offspring
- repeated each test hundreds of times.

Now try this

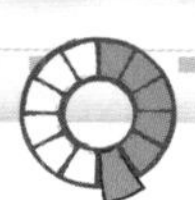

Give **two** reasons that explain why the importance of Mendel's work was not recognised until after his death.
(2 marks)

Fossils

The **fossil record** provides evidence for the theory of **evolution** by **natural selection**.

Fossil formation

Fossils are found in rocks. They are the remains of organisms from millions of years ago. Fossils form in different ways, including:

1. when **traces** of organisms are preserved, such as burrows, rootlet traces and footprints (like the dinosaur footprints in the photo)
2. when parts of the organism, such as bones and teeth, are replaced by minerals as they decay
3. when soft tissues, such as muscles and leaves, do not decay because one or more of the conditions needed is missing.

These trace fossils in Arizona are dinosaur footprints. They were laid down in soft mud around 200 million years ago.

Decay does not happen in the absence of oxygen or water, or at low temperatures.

Worked example

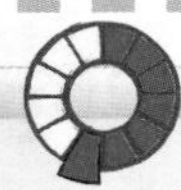

Describe two reasons why scientists cannot be certain about how life on Earth began. **(2 marks)**

Many early forms of life had soft bodies, so they left few traces behind for scientists to find. In addition, most of the traces that were formed have been destroyed by later geological activity.

The fossil record is incomplete because of factors like this. Other reasons include:

- dead organisms decaying or being eaten before fossilisation can begin
- fossils of smaller, more delicate organisms less likely to survive or be discovered.

Extinction

Extinction happens when there are no living individuals of a species left.

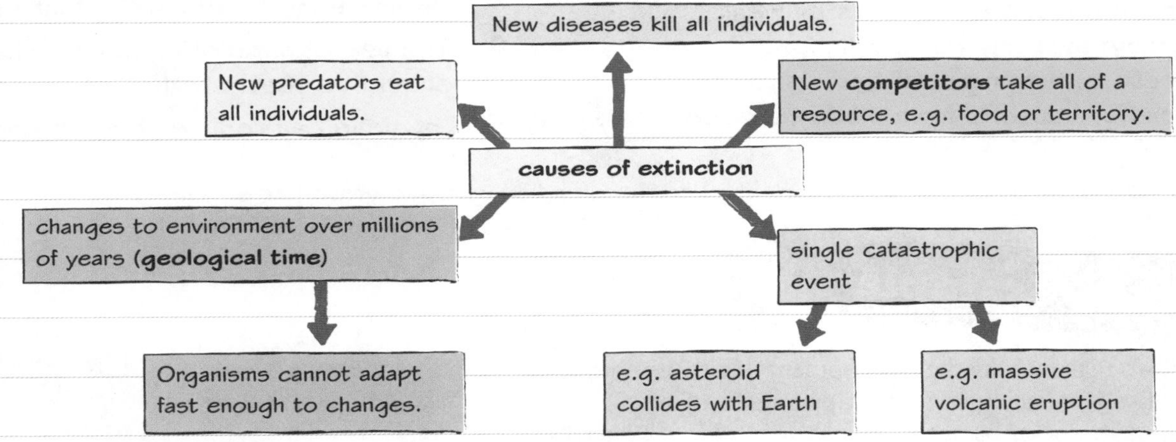

Now try this

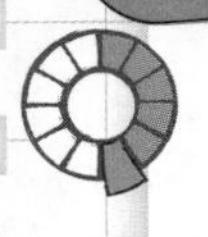

1 Give **one** reason why we have more evidence of how vertebrates (animals with bones) evolved than how soft-bodied animals evolved. **(1 mark)**

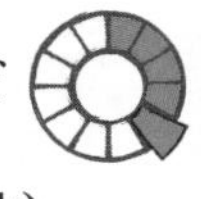

2 State what is meant by a fossil. **(2 marks)**

Resistant bacteria

Resistant strains

Bacterial strains resistant to an antibiotic can spread through a population of bacteria.

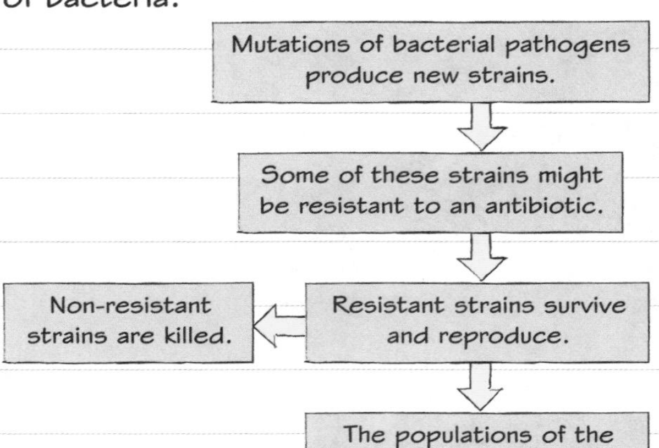

Multiple antibiotic resistance

The development of **antibiotic-resistant** strains of bacteria involves **natural selection**:

- The presence of an antibiotic provides an environmental condition that selects against bacteria that are killed or prevented from reproducing by the antibiotic.

Bacteria can evolve rapidly because they have a high rate of reproduction. A strain of bacteria that is resistant to one particular antibiotic may also develop resistance to another antibiotic:

- Mutations may produce individuals resistant to the second antibiotic.
- These individuals survive in both antibiotics.
- The population of this new strain increases.

MRSA

MRSA stands for meticillin-resistant *Staphylococcus aureus*:

- **Meticillin** (methicillin) is an antibiotic.
- ***S. aureus*** is a type of bacterium.

S. aureus is a common cause of skin and respiratory infections and food poisoning.

Meticillin is the antibiotic that used to be prescribed to treat *S. aureus* infections. It is not effective against MRSA. This causes problems in hospitals because:

- Patients may have wounds or weak immune systems due to illness.
- Doctors, nursing staff and visitors may transfer bacteria from place to place.

Reducing antibiotic resistance

We must reduce the development and spread of antibiotic-resistant strains because:

- people may not be immune to the bacteria
- there may be no effective treatment.

To reduce the rate that these strains form:

1. Doctors should not prescribe antibiotics to treat viral or non-serious bacterial infections.
2. Patients should finish antibiotic courses so all bacteria are killed, leaving none to form resistant strains.
3. The use of antibiotics in agriculture should be restricted.

Careful hygiene can reduce their spread.

Antibiotics are used on farms to prevent infections and to increase growth rate.

Worked example

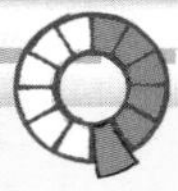

Give two reasons why the development of new antibiotics is not likely to keep up with the appearance of new resistant strains of bacteria. **(2 marks)**

It is expensive to develop new antibiotics, and the process takes a long time.

Now try this

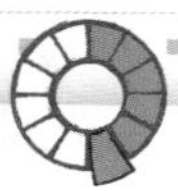

Give **two** ways in which the rate of development of antibiotic-resistant strains of bacteria can be reduced. **(2 marks)**

Classification

Classification involves grouping things in a systematic way according to their similarities.

Carl Linnaeus (1707–1778)

Linnaeus was a Swedish scientist. His **Linnaean system** classifies organisms into groups using their structure and characteristics:

- Kingdoms are the largest groups.
- **Species** are the smallest groups.

A mnemonic like this can help you remember the groups in the right order: **K**ate's **P**oor **C**at **O**nly **F**eels **G**ood **S**ometimes.

Kingdom	Phylum	Class	Order	Family	Genus	Species
group of similar phyla, e.g. Animalia	group of similar classes, e.g. Chordata	group of similar orders, e.g. Mammalia	group of similar families, e.g. Carnivora	group of similar genera, e.g. Canidae (dog family)	group of similar species, e.g. *Canis*	organisms that have most characteristics in common, e.g. *Canis lupus* (domesticated dog)

The binomial system

Every species has a unique **binomial** name. For example, the African lion is called *Panthera leo*.

genus name – shared with very similar species

species name – unique to African lions

Closely related organisms have the same genus name. For example:

- Leopards are *Panthera pardus*.
- Tigers are *Panthera tigris*.

Cheetahs are *Acinonyx jubatus*, showing that they are in a different genus and not closely related to lions, leopards and tigers.

Features of binomial names

Note that:

- the genus name starts with a capital letter but the species name does not
- typed binomial names are in *italics*
- written binomial names are underlined
- the genus name can be abbreviated to its first letter, e.g. *P. leo* for *Panthera leo*.

Binomial names are useful because:

- other people know exactly which species you mean
- you can see from the genus which species are very closely related (but modern DNA analysis can sometimes show that they are not).

Worked example

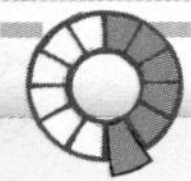

Explain why viruses are not in a kingdom. **(3 marks)**

Viruses do not have all the features of living organisms. Although they reproduce, they must use cells to do this. Viruses do not show other life processes such as respiration or growth.

You can revise viral diseases on page 34.

You may recall 'Mrs Gren' for life processes:

- **M**ovement
- **R**espiration
- **S**ensitivity
- **G**rowth
- **R**eproduction
- **E**xcretion
- **N**utrition

Now try this

The common frog and viper (snake) are in the phylum Chordata. The viper and chameleon are in the class Reptilia. Explain whether the frog or the chameleon shares more characteristics with the viper. **(2 marks)**

Evolutionary trees

Developments in biology have had impacts on **classification systems**.

Three-domain system

Carl Woese proposed a **three-domain system** in 1977. This groups living things into three **domains** using evidence from chemical analysis:

1. **Eukaryota**, which includes plants, animals, fungi and protists
2. **Bacteria** – true bacteria and cyanobacteria (which can photosynthesise)
3. **Archaea** – primitive bacteria that usually live in extreme environments such as very hot or salty water.

Developing systems

The Linnaean system developed over time from two **kingdoms** (plants and animals) to six kingdoms or more. Simple organisms are more difficult to classify than complex ones.

New classification systems have been proposed because of improvements in:
- **microscopes**, increasing understanding of the internal structures of living organisms
- **chemical techniques**, increasing understanding of biochemical processes.

Evolutionary trees

An **evolutionary tree** is a way of showing the evolutionary relationships between species:
- Fossil data are used to work out relationships between extinct species.
- Modern classification data, including the results of DNA analysis, are used to work out relationships between living species.

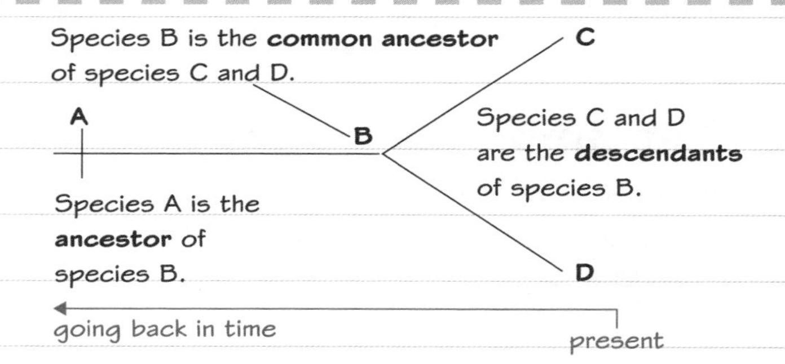

Worked example

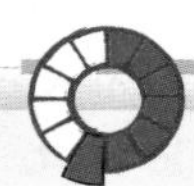

The diagram shows an evolutionary tree.
(a) Describe what the point labelled **X** shows. **(2 marks)**

X shows the last time at which the most recent common ancestor of humans and chimpanzees lived. This common ancestor is now extinct.

(b) Describe what the evolutionary tree shows about cows, whales and pigs. **(2 marks)**

All three share a common ancestor. Cows and whales also share a more recent common ancestor. Both common ancestors are now extinct.

All six types of animal share a common ancestor in the distant past.

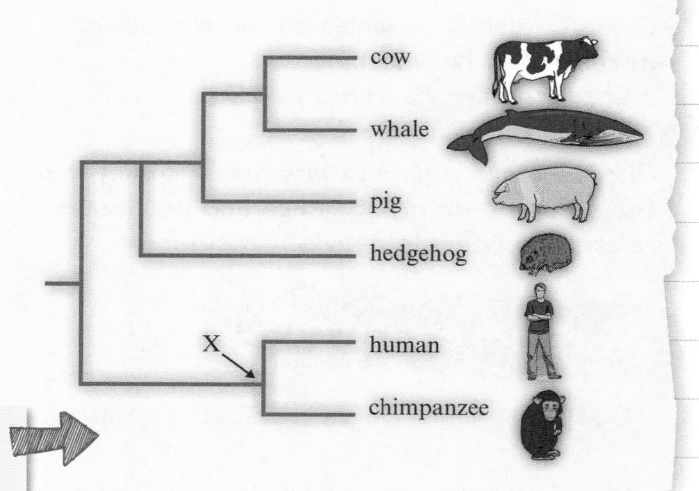

Now try this

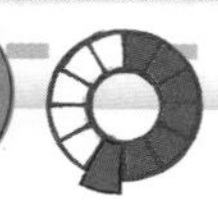

Look at the evolutionary tree in the worked example.
(a) Describe what the tree shows about how similar humans, cows and pigs are. **(2 marks)**
(b) Identify the pair of animals that shares the most recent common ancestor: cow and pig, or pig and hedgehog. Give a reason for your answer. **(2 marks)**

Extended response – Inheritance, variation and evolution

There will be at least one 6-mark question on your exam paper. For these questions, you will need to think scientifically and structure your answer logically, showing how the points you make are related to each other.

You can revise the topics for this question, which is about variation and evolution, on pages 74, 78 and 82.

Worked example

The Colorado beetle is a major insect pest. Each female beetle can lay up to 800 eggs, which hatch after one or two weeks. The larvae become adults quickly enough for three or more generations per year to be produced. Natural predators cannot keep Colorado beetle populations small enough to avoid significant crop damage, so farmers use insecticides. Different populations of Colorado beetle have developed resistance to different insecticides. The species as a whole has developed resistance to over 50 different insecticides.

Explain how the presence of populations of resistant beetles helps to support the theory of evolution by natural selection. **(6 marks)**

The theory describes how evolution happens through natural selection of variants, producing phenotypes that are best suited to their environment. The Colorado beetles best suited to their environment are the ones that are resistant to the insecticides a farmer uses.

Before an insecticide is first sprayed on a field, the population of beetles there may contain some individuals that are resistant to the insecticide. The alleles that cause this are produced by mutation. When the insecticide is used, the resistant beetles survive to reproduce while the other beetles die.

The resistant beetles pass the alleles for resistance to their offspring. The characteristics of the offspring are more likely to include resistance to the insecticide, so the offspring survive to reproduce. Beetles that are not resistant die out. Over time, the population of beetles becomes resistant to the insecticide.

This information shows that populations of Colorado beetles can be large, and that their generation time is relatively short.

This situation is like the development of antibiotic-resistant strains of bacteria.

Command word: Explain

You need to make something clear, or state the reasons for something happening.

This part of the answer outlines the theory and links it to the specific example of the Colorado beetle.

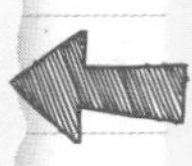

Individual organisms of a species show a wide range of variation for a characteristic. The individuals with characteristics most suited to the environment are more likely to survive to breed successfully.

The characteristics that have allowed these individuals to survive are then passed on to the next generation.

Now try this

The inherited characteristics of a population can change over time because of natural selection or selective breeding. Compare these two processes. **(6 marks)**

You can revise Variation and evolution on page 74 and Selective breeding on page 75.

Ecosystems

Levels of organisation

Ecosystems are organised at different levels.

organism	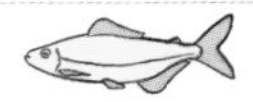	a single living individual
population		all the **organisms** of the same species in a habitat
community		the **populations** of all the different species in a habitat
ecosystem		the interaction of a **community** of living organisms with the non-living parts of their environment

A habitat is where an organism lives. The **environment** is all the conditions that surround an organism in its habitat. These are due to **abiotic** (non-living) factors and **biotic** (living) factors.

Abiotic factors

Abiotic (non-living) factors which can affect a community include:

- ✓ light intensity
- ✓ temperature
- ✓ moisture levels
- ✓ pH and mineral content of the soil
- ✓ intensity and direction of the wind.

In addition:

- oxygen levels are important to aquatic animals such as fish
- carbon dioxide levels are important for plants (as it is used in **photosynthesis**).

Biotic factors

Biotic (living) factors which can affect a community include:

- ✓ the availability of food
- ✓ whether new predators arrive
- ✓ whether new pathogens arrive
- ✓ **competition** between individuals of the same or different species.

Animals often compete with each other for:

- food
- mates
- territory.

One species may outcompete another species, leaving too few individuals to breed.

Worked example

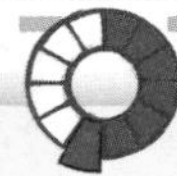

Describe three abiotic factors that plants in a community or habitat often compete for. **(3 marks)**

Plants often compete with each other for light. They also often compete for water and mineral ions in the soil.

Plants also compete for space:

- below ground for their roots
- above ground for their shoots and leaves.

Light and water are needed for photosynthesis. Nitrate ions are needed to make proteins.

You can revise photosynthesis on page 44.

Now try this

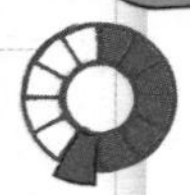

1 Describe the different levels of organisation in an ecosystem, from individual organisms to the whole ecosystem. **(4 marks)**

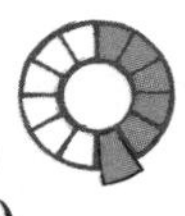

2 The ground underneath the trees in a woodland is mostly bare. Suggest a reason why very few plants can grow there. **(1 mark)**

Interdependence

If one species is removed from a community, it can affect the remaining species. This is called **interdependence**.

Stable communities

Organisms need materials from their surroundings and from other organisms so that they can survive and reproduce. Each species in a community depends on other species for resources, including food and shelter.

In a **stable community**:

- all the species and environmental factors are in balance, so
- the populations stay fairly constant in size.

Plant reproduction

In plants, male **gametes** (pollen cells) are transferred to female gametes (egg cells) by **pollination**. Some species of plants rely on the wind to carry their pollen from one plant to another. Other species rely on bees or other insects to do this.

Many species of plants rely on the wind to disperse their seeds far from the parent plants. Other species rely on animals to do this, for example when they eat fruits or carry seeds on their fur.

An example community

The diagram shows an interdependent community of plants and animals in a woodland habitat.

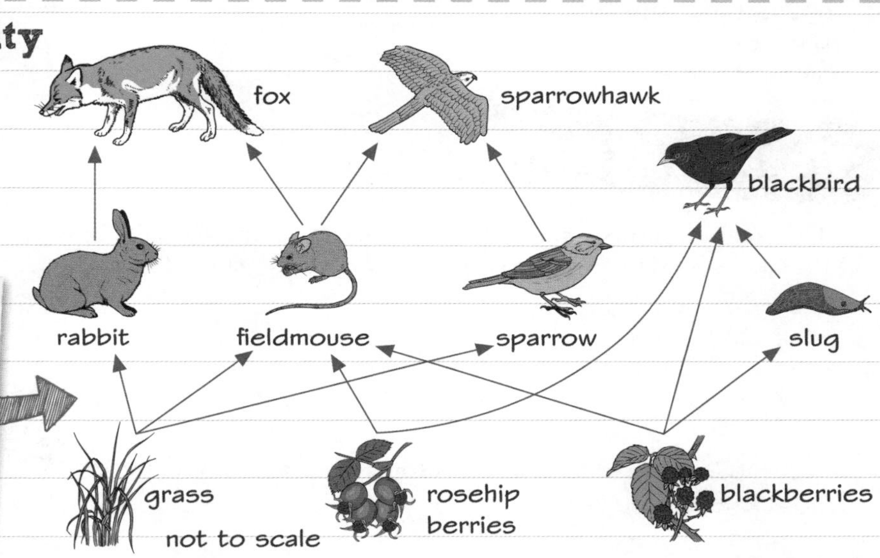

The arrows point from a food source to the animal or animals that eat it. This arrow shows that rabbits eat grass. Notice that the plants are eaten by two or more species, and some animals have more than one food source.

Worked example

Refer to the community in the diagram.

(a) Identify the food eaten by fieldmice. **(1 mark)**

The fieldmice eat grass, rosehip berries and blackberries.

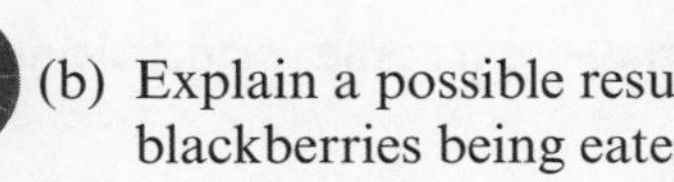

(b) Explain a possible result of all the blackberries being eaten. **(3 marks)**

The fieldmice would need to eat more grass. This would reduce the amount of food available for the rabbits and sparrows. All three species compete for this resource, so the population of one or more of them would fall if there was not enough grass.

Foxes and sparrowhawks eat fieldmice, and also rabbits or sparrows. Their populations would also fall.

Now try this

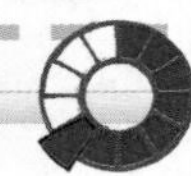

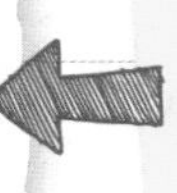

Sparrowhawks are protected by law. However, a hunter illegally shoots these birds in the woodland. Explain **two** possible results of this new predator killing all the sparrowhawks. **(4 marks)**

Refer to the community in the diagram. What do sparrowhawks eat?

Adaptation

Organisms have features that let them survive in the conditions in which they normally live. These **adaptations** may be structural, behavioural or functional.

Cold environments

Animals in cold environments are **adapted** to reduce heat loss, and to move over snow.

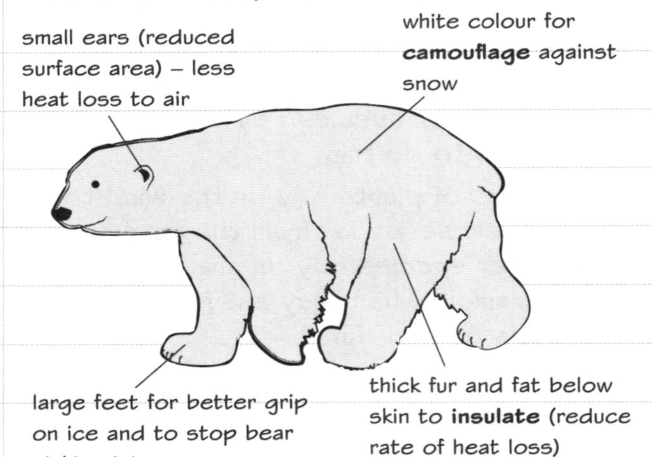

Hot, dry environments

Animals in hot, dry environments are adapted to reduce heat gain and to maintain their water balance. For example, the large ears of African elephants transfer body heat to the air quickly. Camels have:

- thick fur at the top of the body that insulates against heat from the Sun, but thin fur elsewhere to allow heat loss
- the ability to tolerate high body temperatures and low body water content
- a hump of fat, which acts as a food store without insulating the rest of the body against heat loss.

Worked example

The diagram shows some features of a cactus plant. Explain how the cactus is adapted to survive in a dry environment. **(3 marks)**

The large root system allows the cactus to reach water deep underground. The thick fleshy body contains tissues that allow the cactus to store water. The green body allows the cactus to photosynthesise even though it has no leaves.

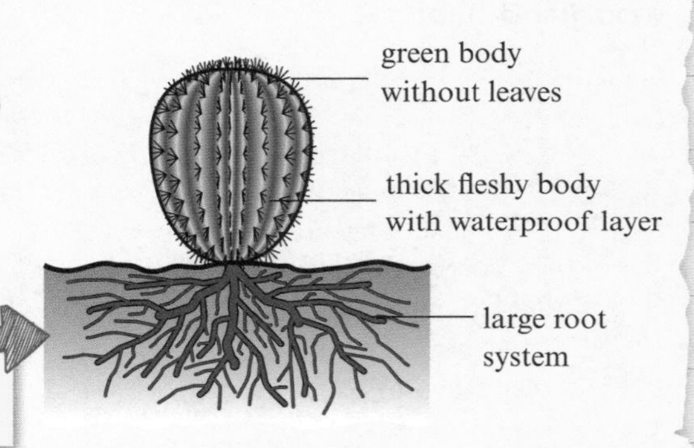

The leaves are adapted to form spikes and thorns to deter animals. You can revise plant defences on page 42.

Extreme environments

Some environments have extreme conditions. These include:

- high temperatures, e.g. volcanic springs
- high levels of salt, e.g. salt marshes
- high pressures, e.g. under the deep ocean.

Extremophiles

Extremophiles are organisms that live in very extreme environments. They have adaptations to let them survive in these environments.

Many extremophiles are microorganisms. For example, certain bacteria live in deep sea vents, where it is very hot, salty and under pressure.

Now try this

1 The Arctic hare lives where there is deep snow for many months, but the snow melts in summer. Explain why the hare has white fur in winter, but brown fur in the summer. **(2 marks)**

2 Explain **two** ways in which the polar bear is adapted to life in a cold environment. **(4 marks)**

A polar bear is shown on this page.

Food chains

Feeding relationships within a community can be represented by **food chains**.

Producers and consumers

In a food chain, each arrow points from the organism being eaten to the organism that eats it. In the food chain in the diagram:

- rabbits eat grass
- foxes eat rabbits.

In general:

- producers are eaten by **primary consumers**
- primary consumers are eaten by **secondary consumers**
- secondary consumers are eaten by **tertiary consumers**.

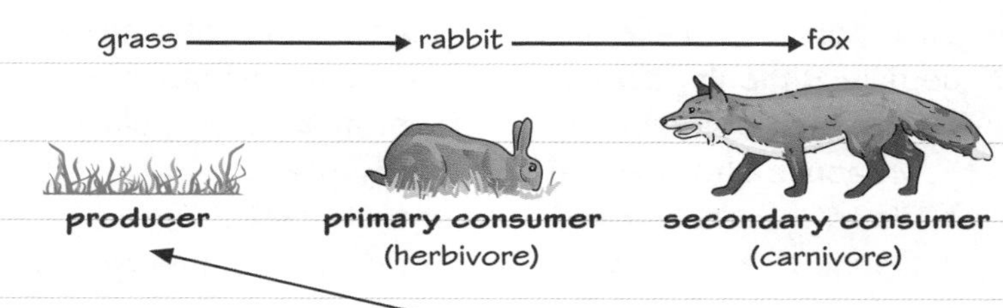

All food chains begin with a **producer**, an organism that synthesises molecules. This is usually a green plant or alga. These make glucose by photosynthesis and produce **biomass** for life on Earth.

Predator–prey cycles

Predators are consumers that kill and eat other animals. **Prey** are the animals that are eaten.

In a **stable community**, the numbers of predators and prey rise and fall in cycles. In each cycle:

1. The prey population rises if there is plenty of food and few predators. More prey survive to reproduce.
2. As the prey population rises, there is more food for the predators. More of them survive to reproduce.
3. As the predator population rises, they eat a greater proportion of prey, so the prey population falls.
4. As the prey population falls, there is less food for the predators, so their population falls too.

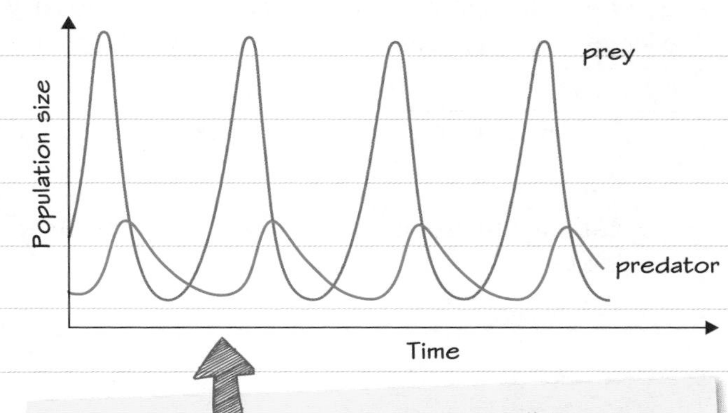

If the population size of the prey fell to zero, there would be no food for the predators. They would all starve and die unless they could find another food source.

The cycle repeats because the number of prey increases again:

- The remaining prey have more food.
- There are fewer predators to eat them.

Worked example

The graph shows the population sizes of zebra and lions in a community. Explain the difference in the sizes of the two populations. **(3 marks)**

Lions are predators and zebra are their prey. A lion must eat several zebra each year to survive. This means that, although the numbers of lions and zebras change in cycles, there are fewer lions than zebra.

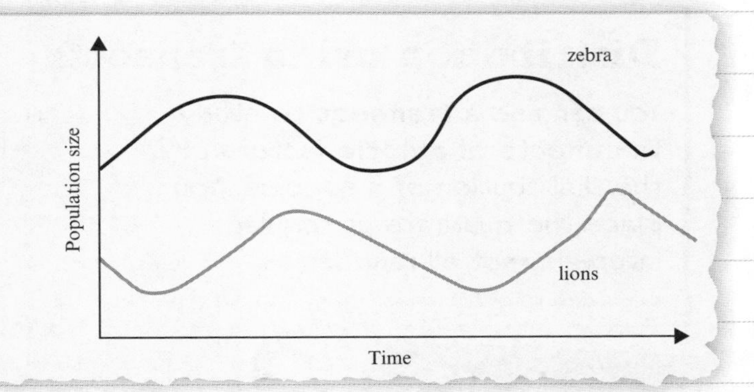

Now try this

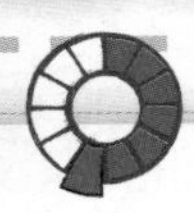

Look at the graphs on this page.

Explain why the population of a prey species decreases after the population of its predator increases. **(2 marks)**

Fieldwork techniques

Many methods are used to find the **abundance** and **distribution** of species in an ecosystem.

Abundance using quadrats

The most accurate way to find the size of a population is to count every individual. This is usually difficult, so samples are studied instead. A study place is sampled with a **quadrat** – a frame with a known area. To estimate a population size, you:

- Measure the area of the study place.
- Place quadrats randomly in the place.
- Count the number of study organisms inside each quadrat.

Quadrats are only useful for plants or very slow-moving animals. Random sampling is important because the organisms may not be evenly distributed in a study place.

Worked example

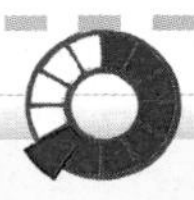

A rectangular school field measures 10 m × 20 m. Ten 0.25 m² quadrats are placed randomly. The number of daisy plants in each quadrat is counted. The table shows the results.

2	4	3	0	2	1	1	4	0	0

(a) Calculate the mean number of daisies in a quadrat. **(1 mark)**

mean number $= \frac{17}{10} = 1.7$

(b) Estimate the total population of daisy plants in the school field. **(2 marks)**

total population $= 1.7 \times \frac{200}{0.25} = 1360$

area of school field $= 10 \times 20 = 200\ m^2$

Small quadrats are often 0.5 m × 0.5 m, giving an area of 0.25 m².

Maths skills: Calculating a mean

To calculate a mean:

- Add the values together to get a total.
- Divide the total by the number of values.

Maths skills: Mode and median

You can find these when you put the values in ascending order, e.g. for the daisy data:

0	0	0	1	1	2	2	3	4	4

The **mode** is the most frequently occurring value. It is 0 in these data (it occurs three times).

The **median** is the middle value in a range. If there is an even number of values (as here) you calculate the mean of the middle two values. This is $\frac{(1 + 2)}{2} = 1.5$ here.

Distribution using transects

You can use a **transect** to study the effects of **abiotic** factors on the distribution of a species. You place the quadrats at regular intervals, not at random. You can find more details about this on page 91.

1 m × 1 m quadrats placed at regular intervals along the transect

transect line – e.g. tape measure placed along the ground

Now try this

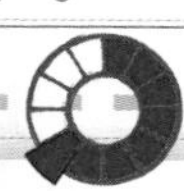

Ten 1 m² quadrats were placed randomly on a 100 m by 200 m field. The number of dandelions recorded were: 20, 6, 33, 0, 26, 21, 18, 7, 2, and 9. Estimate the total number of dandelions on the field. **(3 marks)**

Field investigations

You can use quadrats and transects to investigate the relationship between organisms and their environment in the field.

Core practical

Measuring a population size in a habitat

Aim

to estimate the population of a plant species using random sampling

Method

1 Use two tape measures to mark out and measure a large area of a field.
2 Record the number of dandelions inside at least 10 randomly placed 25 cm × 25 cm quadrats. Record your results in a suitable table.

Analysis

Use your results, and the marked area of the field, to estimate the population size.

Dandelion plants are easily recognised.

You can revise this field technique on page 90. The area of a 25 cm quadrat is 0.0625 m^2 (0.25 m × 0.25 m).

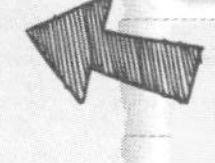

The estimated population size will be more accurate if you share results with other groups.

Investigating the effect of a factor

Aim

to use a transect line to investigate the effect of light intensity

Method

1 Lay a tape measure in a straight line from the base of a tree to open ground.
2 Place a quadrat on the 0 m mark.
3 Count the number of dandelion plants inside the quadrat. Measure and record the light intensity using a light meter.
4 Repeat step 3 at regular intervals on the line.
5 Record your results in a suitable table.

You can revise how to do this on page 90.

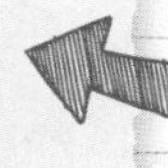

You could also investigate other abiotic factors such as soil pH and temperature.

Analysis

Plot a graph to show the results. This one shows some example results.

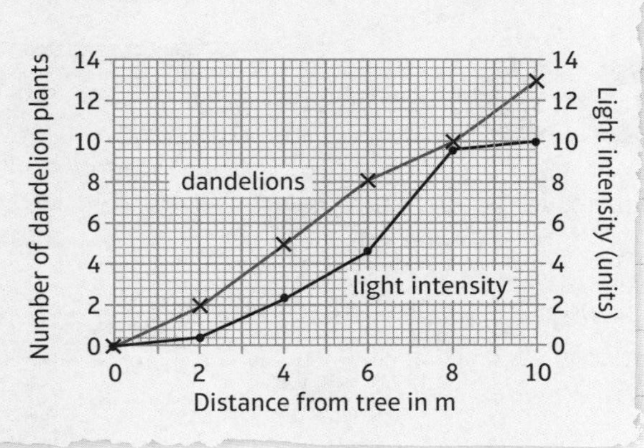

Take care not to lay your transect in dangerous places. The interval chosen is likely to differ between investigations. If it is:

- small, the number of samples will be large and take time to study
- large, you may miss important detail from your investigation.

Remember to count plants, not flowers.

Now try this

(a) Describe how the number of dandelion plants changes with distance from the tree. **(1 mark)**
(b) Suggest an explanation for the distribution observed. **(3 marks)**

Cycling materials

The water cycle

The **water cycle** provides land plants and animals with fresh water.

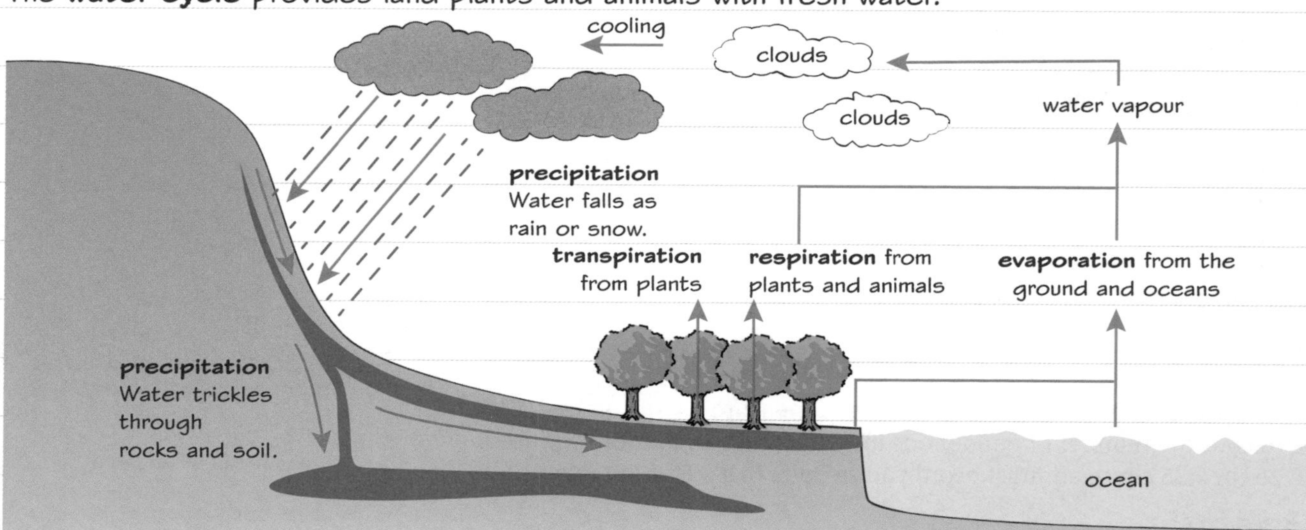

The carbon cycle

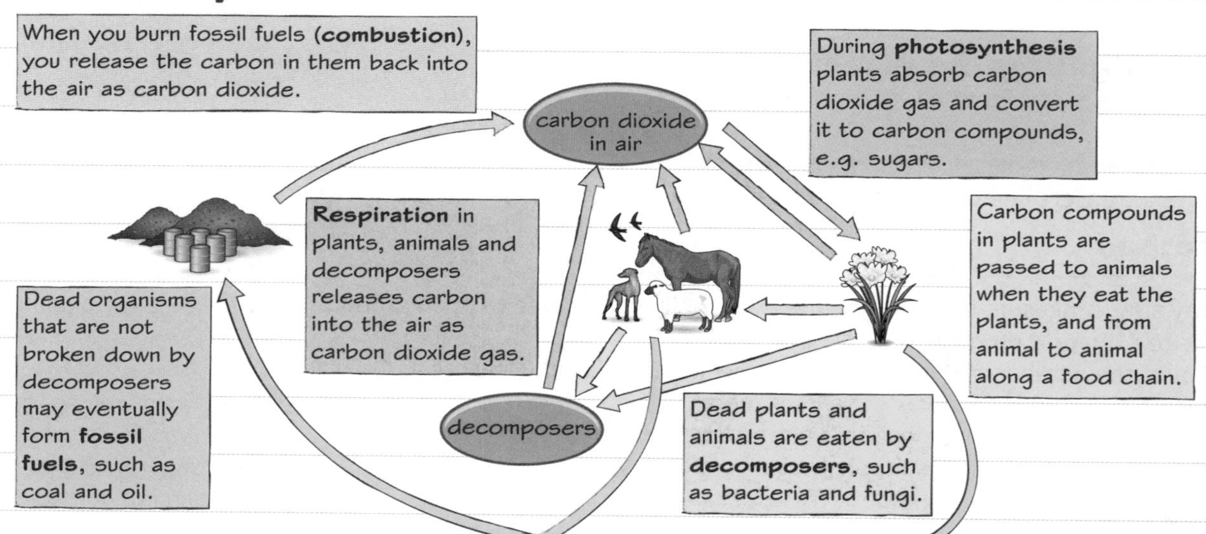

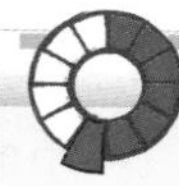

Worked example

Explain the role of microorganisms in the carbon cycle. **(3 marks)**

Decomposers are microorganisms that feed on dead organisms and their waste. They digest this material, and use the nutrients to grow and reproduce. They release carbon dioxide to the atmosphere as they respire.

The decay process is often started when worms, beetles and other detritus feeders eat dead organisms.

Decomposers include bacteria and fungi. They also return mineral ions to the soil. Nitrate ions are needed by plants to make amino acids.

Now try this

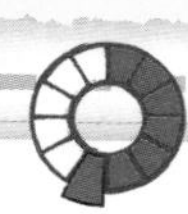

Identify steps in the carbon cycle that:

(a) return carbon dioxide to the atmosphere **(2 marks)**

(b) remove carbon dioxide from the atmosphere. **(1 mark)**

Decomposition

Decomposition is the breakdown or **decay** of biological materials.

Rate of decay

Factors affecting the rate of decay include:

1. **Temperature**
   Warm temperatures increase the rate of decay because they increase the rate of chemical reactions in microorganisms.
2. **Water content**
   Decay happens faster in moist conditions because microorganisms need water for their cell processes.
3. **Oxygen availability**
   Most decomposers use **aerobic respiration**. Decay is faster in the presence of oxygen because they can respire.

Biogas generators

Some decomposers use **anaerobic respiration**. Anaerobic decay produces **methane** gas. Methane is used as a fuel for cooking and heating. A **biogas generator** uses waste from animals to produce large volumes of methane.

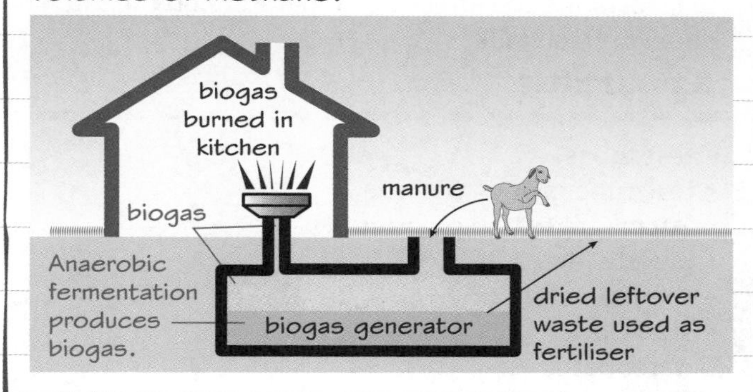

Making compost

Gardeners produce **compost** from kitchen and garden waste. This is mainly decayed plant material, and it acts as a natural fertiliser for growing garden plants.

Farmers produce compost and **manure** (which comes from animal waste) on larger scales. They use this material as a natural fertiliser for growing crops.

Worked example

A student investigates the decay of grass clippings. She sets up three insulated flasks. She fills one with damp grass, one with dry grass, and one with damp grass and disinfectant.
She places the flasks in the same room without lids. The student measures and records the mass of the green clippings. The table shows her results.

Conditions	Mass of green clippings (g) Day 0	Day 6	Rate of decay (g/day)
damp	50	5	7.5
dry	50	22	4.7
disinfectant	50	44	1.0

(a) Complete the table by calculating the rate of decay in each set of conditions. **(1 mark)**

(b) Explain the conditions in which the grass decayed most quickly. **(2 marks)**

The damp grass clippings decayed most quickly. This was because water and oxygen were available, and the flask maintained a warm temperature. These are the conditions in which decay microorganisms grow best.

Dry clippings decay slowly because there is insufficient water for microorganisms to grow well. Disinfectant kills microorganisms, so decay happens very slowly in its presence.

Now try this

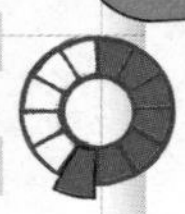

1 Explain why compost heaps produce compost most quickly in sunny places. **(2 marks)**

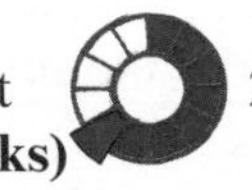

2 Suggest an explanation for why compost forms faster if it is turned over regularly. **(2 marks)**

Investigating decay

You can investigate the effect of temperature on the decay of milk by measuring pH change.

Core practical

Investigating the rate of decay

Aim

to investigate the effect of temperature on the rate of decay of fresh milk

Natural decay of milk produces lactic acid, which causes the pH to fall. It may take several days for milk to decay or 'go off'. Lipase is used to model the decay of milk. The process is quicker and produces fatty acids, which also cause the pH to fall.

Apparatus

- test tubes
- test-tube rack
- 10 cm^3 graduated pipettes
- pipette filler
- thermometer
- stop clock
- water baths
- fresh full-fat milk
- lipase solution
- sodium carbonate solution
- cresol red indicator

You may be given graduated plastic syringes instead of graduated pipettes.

Beakers of water from a kettle can be used if thermostatically controlled water baths are not available.

Method

1 To a test tube, add 5 cm^3 of milk, 7 cm^3 of sodium carbonate solution and a few drops of cresol red indicator. Put the thermometer in the test tube.
2 Add about 5 cm^3 of lipase solution to another test tube. Place both test tubes in a water bath and leave them until their temperature remains constant.
3 Add 1 cm^3 of lipase solution to the milk mixture and start the stop clock. Stir until the indicator changes colour and record the time taken.
4 Repeat steps 1 to 3 at different temperatures.

Sodium carbonate makes the mixture alkaline. Cresol red is purple in these conditions.

You could set up more than one test tube at step 1. This allows you to carry out repeats if you have time.

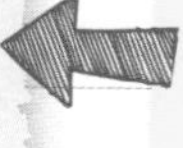

Cresol red is yellow in acidic conditions. You could use phenolphthalein indicator instead. This turns from pink in alkaline conditions to colourless in acidic conditions.

Results

Record your results in a suitable table.

Analysis

Plot a graph to show to show the results:

- time taken for indicator to change colour on the vertical axis
- temperature on the horizontal axis.

You can obtain temperatures below room temperature using an iced water bath. This would extend the range of temperatures investigated.

Now try this

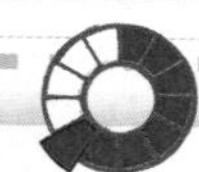

A student carries out an investigation into the effect of temperature on the rate of decay of milk. Lipase solution is used to model the effects of decay bacteria.

(a) Explain why the time taken for the milk to become acidic decreases between 5 °C and 35 °C. **(2 marks)**

(b) Explain why the milk does not become acidic at 55 °C. **(3 marks)**

You can revise the effect of temperature on enzyme activity on page 19.

Waste management

Pollution kills animals and plants, which can lead to a reduction in **biodiversity**.

Pollution

Pollution is caused when a harmful or poisonous substance is released into the environment.

POLLUTION

of the air:
- smoke and gases, e.g. sulfur dioxide, which is part of acid rain

of the land:
- toxic chemicals, e.g. pesticides and herbicides
- dumping waste in landfill sites

of the water:
- sewage
- fertiliser
- toxic chemicals washed in from land

Pesticides are used on crops to kill pests. **Herbicides** are used on crops to kill weeds.

Sewage contains waste water as well as human waste.

Increasing pollution

The human **population** is growing rapidly.

rapid growth in human population + increased standard of living = increasing waste produced → if not handled properly, causes more pollution

uses more land for:
- building
- quarrying for building materials
- farming to produce food
- dumping waste

→ less space for other animals and plants

Worked example

Coal and oil release sulfur dioxide when they are burned. Natural gas releases much less of this toxic gas. In the UK, most sulfur dioxide emissions are from power stations. The graph shows total emissions between 2000 and 2014.

Annual emissions of sulfur dioxide (Year 2000 = 100)

100, 90, 80, 70, 60, 50, 40, 30, 20, 10, 0

2000, 2002, 2004, 2006, 2008, 2010, 2012, 2014

Year

(a) In 2000, electricity generation began to switch from coal and oil to natural gas. Explain how the graph gives evidence of this. **(2 marks)**

Total emissions decreased after 2000. Natural gas releases less sulfur dioxide than coal and oil do, so switching to natural gas should reduce the amount released.

(b) More coal, which is cheaper than oil and gas, was used between 2010 and 2012. State how the graph gives evidence of this. **(1 mark)**

The total emissions were higher in 2010 and 2012 than they were in 2009.

Sulfur dioxide is a major cause of acid rain.

Now try this

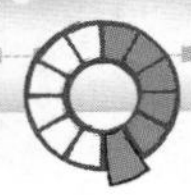

Give **two** reasons why pollution from human activities is increasing. **(2 marks)**

Deforestation

Deforestation is the permanent destruction of forests to make land available for other uses.

Reasons

Large-scale deforestation has already occurred in the UK and other parts of Europe. It is continuing in tropical areas.

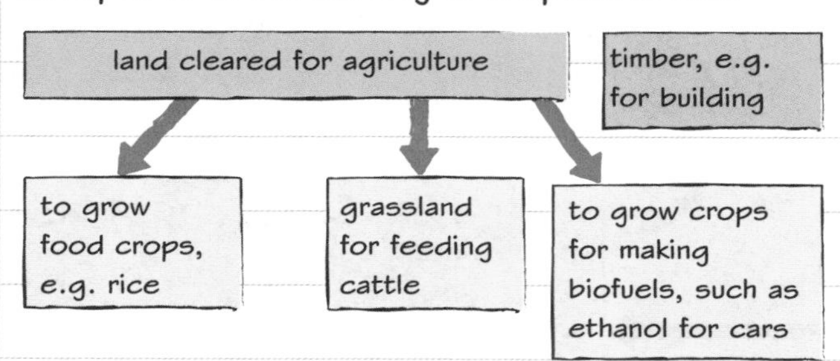

Problems

Deforestation has environmental impacts.

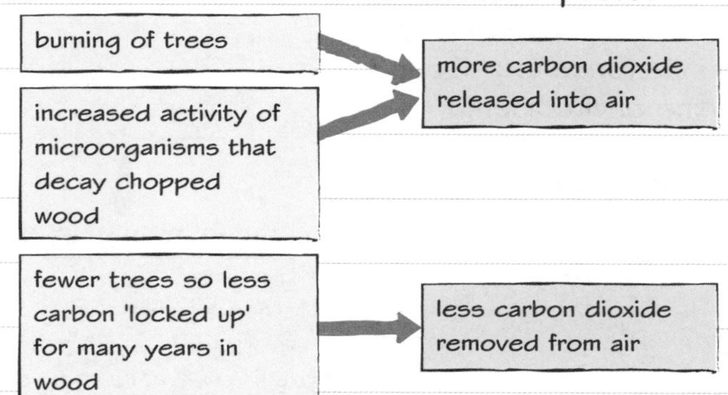

Peat

Peat consists of partly decomposed plant material. It is found in peat bogs and peatland.

Peat is useful:

- as a fuel in some places
- to make garden compost.

Carbon dioxide is released into the atmosphere when peat decays further, or is burned as a fuel. This contributes to **global warming** (see page 97).

The destruction of habitats containing peat reduces the biodiversity there.

Biodiversity

Biodiversity is the variety of all the different species in an ecosystem or on Earth.

A high level of biodiversity keeps ecosystems stable. It reduces **interdependence** between species for:

- ✓ food and shelter
- ✓ maintenance of the physical environment.

Our future as a species depends on us maintaining a good level of biodiversity.

Human activities are reducing biodiversity.

Worked example

Ground-living arthropods include beetles and millipedes. Palm oil comes from the fruit of oil palm trees. In parts of Borneo, areas of tropical rainforest have been cleared so oil palm trees can be planted instead. A study there found between 7 and 11 different arthropod groups in original rainforests. It found 4 or 5 arthropod groups in palm oil plantations.

Explain, using this information, the effect on the level of biodiversity of replacing rainforest with a single type of tree. **(2 marks)**

Replacing different types of rainforest trees with oil palm trees reduces plant diversity.

The study shows a reduced level of biodiversity in terms of the number of groups of arthropods. This is because fewer arthropod groups were found in the plantations than in the original rainforest.

A reduction in the number of arthropod groups will affect other organisms. Further studies might show a reduction in the number of other animals, such as orangutans.

Now try this

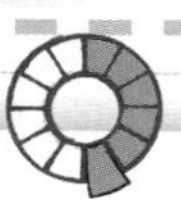

Give **two** reasons why there is large-scale deforestation in tropical areas. **(2 marks)**

Global warming

Greenhouse gases

Greenhouse gases such as methane and carbon dioxide absorb infrared radiation emitted by the Earth's surface, then release it in all directions. This helps to keep the Earth warm. Human activities are increasing the levels of greenhouse gases.

Greenhouse gas	Human activities
methane	farming (rice paddy fields and cattle)
carbon dioxide	combustion of coal, oil, natural gas and peat

Greenhouse effect

Increasing levels of greenhouse gases are causing an enhanced **greenhouse effect**. There is a scientific consensus that this is leading to **global warming** and **climate change**. This consensus is based on:

- thousands of peer-reviewed publications
- systematic reviews of these papers.

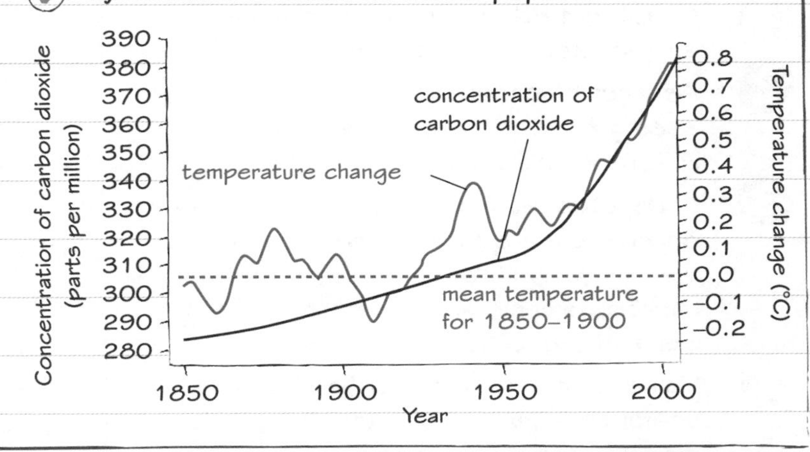

Effects of global warming

An increase in average temperatures of only a few degrees Celsius could have these effects:

- cause big changes in climate → e.g. drought, colder winters, hotter summers, flooding
- cause a rise in sea level → due to ice melting and to warmer sea water (expands as it gets warmer)
- affect species → e.g. change in migration of birds, change in distribution of species (the area they live in)

Worked example

Bluebell plants produce flowers in spring. The date of the first bluebell flower in a particular woodland was recorded between 1996 and 2012. The table shows the results.

Year of study	1996	2000	2004	2008	2012
Date of first bluebell flower	23 April	13 April	14 April	10 April	4 April

(a) Explain one way in which the method used improves the reliability of the results. **(2 marks)**

The data were collected at a single site, rather than from two or more sites. This helps to control factors that differ between sites.

Data from a single species may not lead to a valid conclusion.

(b) Explain whether the data provides evidence for climate change. **(2 marks)**

Overall, the flowering date becomes earlier during the study. This suggests that the climate is changing, for example becoming warmer earlier in the year. However, we do not know what happened in the years between each visit.

Now try this

Suggest how climate change may make it possible for tropical diseases to spread to new areas. **(3 marks)**

Maintaining biodiversity

Many human activities reduce **biodiversity**. Steps have only been taken recently to stop this.

Reasons for positive action

There are programmes to reduce the negative impact of human activities on the environment. These have been put in place by scientists and other concerned people. Reasons include:

1. Moral and ethical reasons: humans should respect other species.
2. Aesthetic reasons: people enjoy seeing the variety of species in different **habitats**.
3. Ecosystem structure: organisms in an ecosystem are **interdependent** (revise this on page 87). For example, decay microorganisms are important in the **carbon cycle** (revise this on page 92).
4. Value: some species are particularly useful to humans. For example, plants may be a source of new medicines (revise this on page 41), and others provide opportunities to make money from tourism.

Some programmes

Programmes to reduce or reverse the decline in biodiversity include:

- protecting and regenerating rare habitats, such as wetlands
- replanting hedgerows in farms
- recycling resources rather than dumping waste in landfill, which reduces activities such as quarrying
- reducing deforestation.

Replanting forests

Advantages of **reforestation** include:

- restores habitat for endangered species
- reduces the effects of soil erosion because tree roots bind soil together
- helps to reduce the overall release of carbon dioxide as the trees photosynthesise.

Worked example

The graph shows the changes in the Asian tiger population.

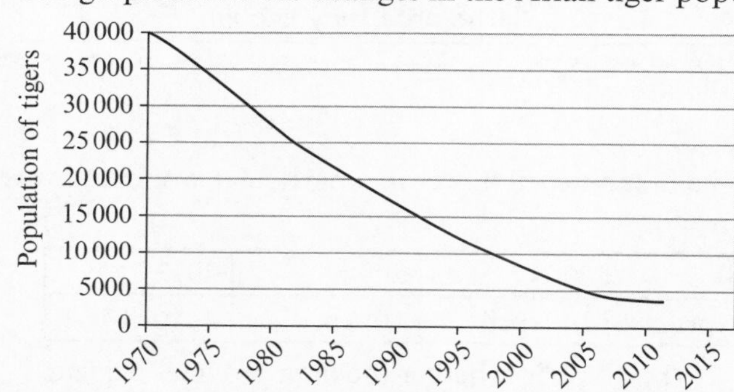

(a) During this time, the human population in tiger regions increased significantly.
Explain how this might be linked to the size of the tiger population. **(2 marks)**

The tiger population decreased as the human population increased. Human activities such as farming and building may be destroying the tigers' habitat. Hunters may be killing tigers.

Extinction happens when no living individuals of a species remain (you can revise causes of this on page 81).

An **endangered species** faces a high risk of extinction in the near future. There are **breeding programmes** for some of these species. Animals or plants are bred over several generations. Reproduction is planned to increase numbers without causing **inbreeding** (you can revise this on page 75).

(b) Scientists may closely monitor the population size of large carnivores like tigers when they assess the biodiversity of an area. Suggest a reason for this. **(2 marks)**

Large carnivores eat other animals, so they may rely on complex food chains. If there are many large carnivores in an area, it suggests that there may be many other species there too.

Now try this

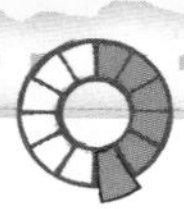

Give **three** ways in which biodiversity can be protected or improved. **(3 marks)**

Trophic levels

Trophic levels are feeding levels along a **food chain**, starting at level 1, the producers.

Pyramids of biomass

Biomass is the mass of material in an organism.

It is often given as the dry mass – the mass of the organism without its water.

A **pyramid of biomass** shows the relative amount of biomass at each trophic level.

You can revise food chains on page 89.

Description	Trophic level
carnivores that eat other carnivores (Apex predators are carnivores with no predators.)	Level 4 tertiary consumers
carnivores that eat herbivores	Level 3 secondary consumers
herbivores that eat plants and algae	Level 2 primary consumers
make their own food	Level 1 producers (plants and algae)

Worked example

The table shows the biomass of organisms in the food chain: lettuce → caterpillar → thrush

Organism	Biomass (g/m²)
lettuce	120
caterpillar	60
thrush	12

(a) Use the data to draw an accurate pyramid of biomass. **(2 marks)**

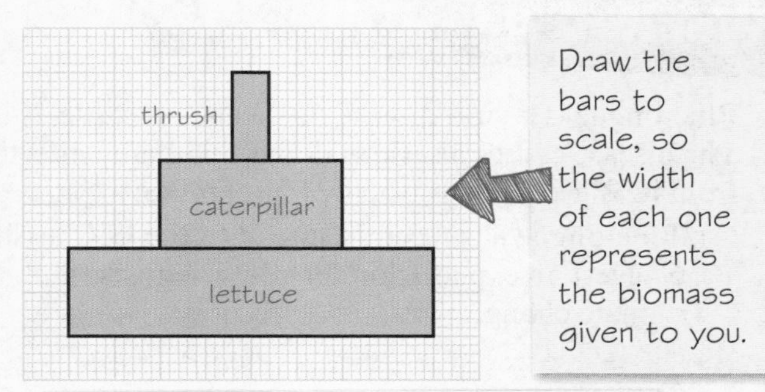

Draw the bars to scale, so the width of each one represents the biomass given to you.

(b) Give one reason why large amounts of glucose are lost at each trophic level. **(1 mark)**

The glucose is used for respiration.

Remember that all living organisms respire, so glucose is used for respiration in plants **and** animals.

(c) Explain the shape of the pyramid of biomass. **(3 marks)**

Biomass is lost between one trophic level and the next. This is because the caterpillars and thrushes cannot digest and absorb all the material they eat. Some of the biomass eaten by the caterpillars is egested as faeces, so is not eaten by the thrushes.

About 1% of the energy from light received by plants and algae is transferred for photosynthesis.

In general, only about 10% of the biomass in one trophic level is passed to the level above.

Other biomass losses include:

- carbon dioxide and water in respiration
- water and urea in urine.

Maths skills: Efficiency

60 g of caterpillar becomes 12 g of thrush. The **efficiency** is:

$$\frac{12}{60} = \frac{1}{5}$$

Multiply by 100 to obtain a percentage:

$$100 \times \frac{1}{5} = 20\%$$

Numbers

The numbers at each trophic level depend on:

- ✓ the biomass of each individual
- ✓ the total biomass of the trophic level.

For a given total biomass, small organisms will be greater in number than large organisms.

Now try this

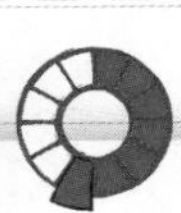

A rabbit eats 500 g of grass. Give **four** reasons why the rabbit will not increase in biomass by 500 g. **(4 marks)**

Food security

Food security is having enough food to feed a population. **Sustainable** methods of food production must be found to feed everyone on Earth.

Increasing birth rate

The **birth rate** is increasing in parts of the world. This affects food security and the environment.

Increasing **human population** means more food is needed. → The increasing demand for meat and fish means more land is used for **animal farming** and a greater impact on wild fish populations. → Movement of people and goods introduces new pests and pathogens to areas, damaging local crops and animals. → Increased waste is produced, causing more pollution.

Worked example

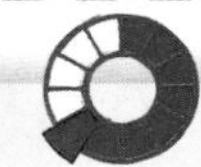

Bluetongue is a viral disease of sheep and cattle. It is spread by insects. The disease causes a serious illness and many infected animals will die from it. Bluetongue is usually found in warm parts of southern Europe.

(a) Bluetongue has recently spread to the UK, as shown on the map. Suggest an explanation for why scientists think this may be due to climate change. **(2 marks)**

The insect vector that spreads bluetongue may need a warm climate to survive. If climate change is making the UK warmer, it is likely that these insects can now survive in the UK.

(b) Give one reason why farmers in the UK are worried about bluetongue. **(1 mark)**

The disease will make farm animals ill or even kill them. This would reduce the food supply.

When new diseases or pests spread to an area, the existing plants or animals may not have resistance against them. This can seriously affect farming, reducing the yield and quality of the food produced, and the farmers' profits.

Other factors affecting food security

Conflicts can also reduce the availability of fresh water for drinking and for watering crops.

Reduced food supply:

- Wars and other conflicts make it difficult for farmers to grow, process and transport food.
- increased cost of fuel, fertilisers, machinery, seed, animal feed
- changing diets in developed countries:
  - some food supplied by less well developed countries
- environmental changes, such as reduced rainfall:
  - may cause widespread famine

Now try this

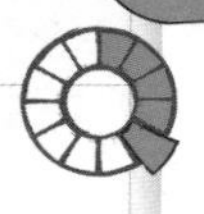

1 State what is meant by food security. **(1 mark)**

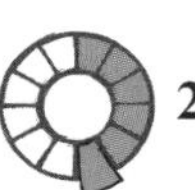

2 Give **three** factors that may threaten food security. **(3 marks)**

Farming techniques

Farmers may use **intensive farming** techniques to increase the yield of food. This type of farming is sometimes called 'factory farming'.

Intensive farming

The efficiency of meat production can be improved by:

- keeping animals in pens or cages to limit their movement
- keeping the surroundings warm.

This reduces energy transfer from the animals to their environment.

Animals may also be given diets high in protein to increase their growth.

Food animals lose energy:
- by moving around
- as heat to the environment.

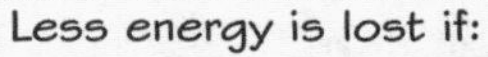

Less energy is lost if:
- movement is restricted, e.g. animals are kept in cages
- the environment is kept warm.

✓ • The animals grow faster and bigger.
✗ • Animals close together are more likely to get ill.
✗ • Heating uses electricity.
✗ • It is unethical to keep animals in stressful or unhealthy conditions.

Worked example

Free-range chickens can move around freely, flap their wings, and behave naturally. Intensively farmed chickens have less room to move around and may be kept in large, heated barns. Free-range chicken is usually more expensive to buy than intensively farmed chicken.
Evaluate the use of these two methods to produce chicken for human consumption. **(5 marks)**

You do not need to give a conclusion unless you are asked to do so.

The free-range chickens may grow more slowly than the intensively farmed chickens. This is because they will lose more energy keeping warm and moving around, so they will use more glucose for respiration. On the other hand, they can behave more naturally than the intensively farmed chickens. People may be happy to pay more money for free-range chicken because they will think it has been produced more ethically. However, some people may not be able to afford free-range chicken.

Plant diet vs animal diet

More food is available for humans from the same mass of plant material if we eat the plants, rather than eating meat from animals that eat the plants.

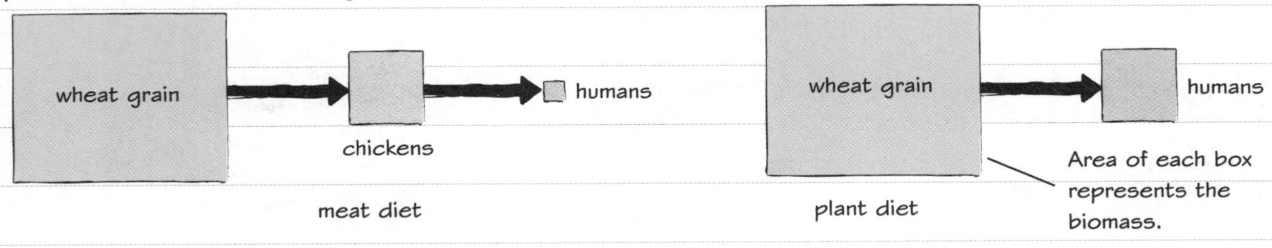

Biomass is lost from one trophic level to the next (you can revise this topic on page 99).

Now try this

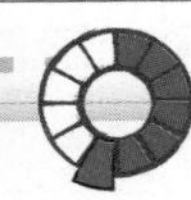

Refer to this food chain. In terms of efficiency of food production, explain why it is better to eat sardines rather than tuna.
microscopic plankton → sardine (small fish) → tuna (large fish) **(2 marks)**

Sustainable fisheries

People catch fish for food. Fish stocks must be maintained at levels that allow breeding to occur, or some species of fish may disappear in some areas.

Catching fish

Fishermen catch many sizes of fish. If there are not enough large fish in the catch, they may also take smaller fish. The rest of the catch is usually thrown back into the water.

It takes several years for fish to mature and become large enough to breed.

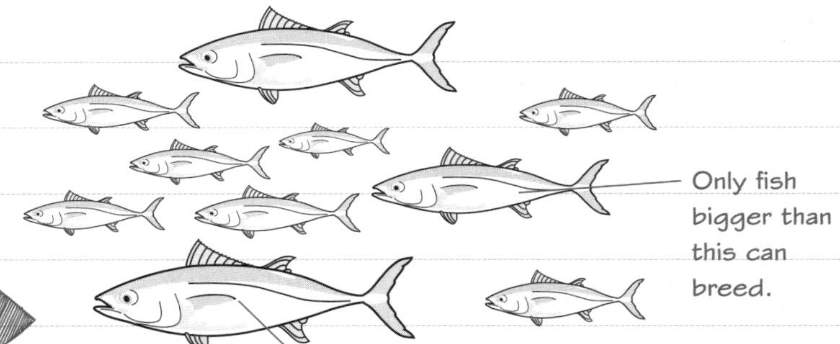

Reasons for declines in fish stocks

If fishing takes all the large fish … → there are no fish left that are big enough to breed … → so no young fish replace the fish taken in fishing … → the fishery collapses (there are not enough fish to make fishing worthwhile).

In a **sustainable fishery**, we could catch enough fish for our needs without affecting the needs of people in the future. The populations of the different species would be stable. However, if we catch too many fish to be sustainable (we overfish), we will have:

- plenty of fish in the short term
- few fish or no fish in the longer term.

Worked example

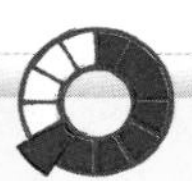

Fish stocks must be conserved to make a fishery sustainable. Fishing quotas and limits on mesh size can achieve this or allow stocks to recover.

(a) Explain what is meant by a fishing quota. **(2 marks)**

A fishing quota is a maximum number or mass of fish that fishermen are allowed to catch. Quotas usually apply to a particular species of fish in an effort to let stocks increase.

(b) Explain how limits on the mesh size of fishing nets can help to conserve fish stocks. **(3 marks)**

The mesh size is limited to a minimum size. This ensures that small fish are not caught. If the mesh limit is larger than the smallest size for a breeding fish, there will always be some fish left in the area that can reproduce.

Another way to allow stocks to recover is to close an area to fishing for several years.

Quotas and other conservation measures are often unpopular with people in the fishing industry because employment is reduced.

Now try this

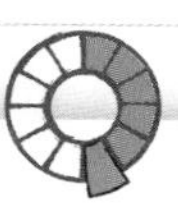

Give **two** ways, other than setting a minimum mesh size, in which fish stocks can be conserved. **(2 marks)**

Biotechnology and food

Mycoprotein

Fusarium is a type of fungus. It is useful for producing **mycoprotein**. This is:

- rich in protein
- suitable for vegetarians.

Fusarium is grown in large amounts in huge containers called fermenters. These are around 50 m high (see diagram). In the fermenter:

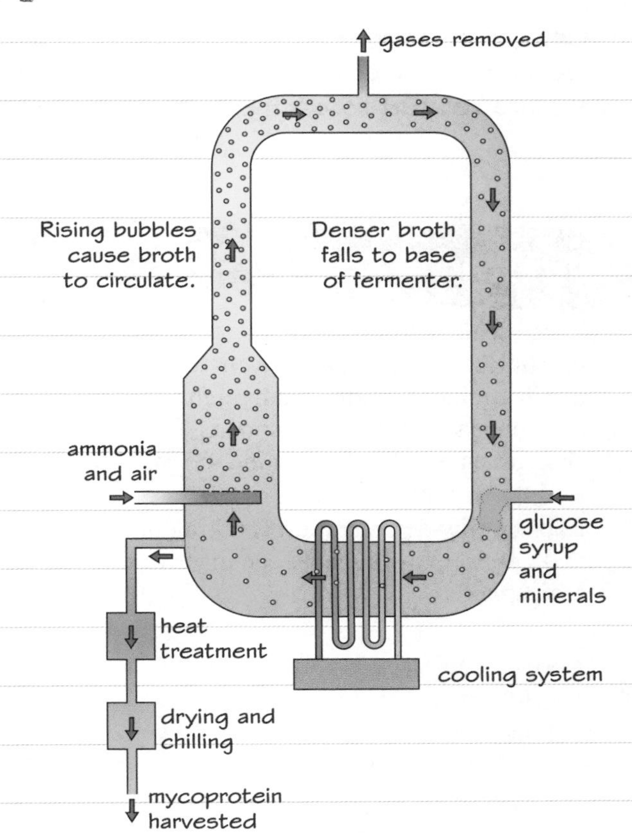

1. Air is added to provide aerobic conditions so the fungal cells can respire.
2. Glucose syrup is added as a source of food for the fungal cells.
3. The culture mixture or 'broth' continually circulates. *Fusarium* is dense and sinks to the bottom.
4. *Fusarium* is harvested and purified. It can be shaped into pieces to imitate the appearance and texture of meat.

Ammonia is added as a source of nitrogen for the *Fusarium* cells to make proteins.

Genetic modification

Genetic engineering is used to produce **genetically modified** (GM) organisms:

- ✓ GM bacteria produce human insulin to treat people with diabetes.
- ✓ GM crops with high yields can provide more food.
- ✓ GM crops such as 'golden rice' can provide improved nutritional value.
- ✓ Making human insulin using GM bacteria is quicker and cheaper than producing it any other way.
- ✗ A few diabetic people react badly to this insulin and need a different form.
- ✓ Golden rice has genes that increase its production of β-carotene, needed to make vitamin A. Eating golden rice could prevent vitamin A deficiency.
- ✗ Golden rice seed costs more than normal rice seed.

You can revise genetic engineering on page 76.

Worked example

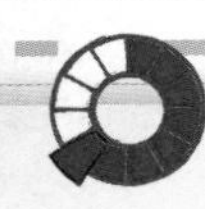

Most mushrooms are grown commercially. Explain why they can be grown on a variety of substances, including wood chips and manure. **(3 marks)**

Mushrooms are decomposers. Like other decomposers, they secrete enzymes into the environment that break down dead plant and animal material. They then absorb the small, soluble molecules formed.

The molecules are absorbed by diffusion.

Now try this

Describe **two** ways in which biotechnology can increase the amount of food for people. **(4 marks)**

Extended response – Ecology

There will be at least one 6-mark question on your exam paper. For these questions, you will need to think scientifically and structure your answer logically, showing how the points you make are related to each other.

You can revise the topics for this question, which is about deforestation and global warming, on pages 96 and 97.

Worked example

Large-scale deforestation has occurred in tropical areas. It has environmental implications, including changing the amounts of some gases in the atmosphere. These changes cause global warming, possibly leading to climate change.

Explain why deforestation is taking place, and describe how it causes changes to the atmosphere. **(6 marks)**

The human population is growing and standards of living are increasing. More land is needed to sustain these changes. Deforestation involves clearing forests to provide more land for human activities.

The land released by deforestation may be used to grow crops for biofuels, such as ethanol for cars. It is also used to provide grazing for cattle, or fields to grow rice. This allows more food to be produced. However, it also leads to an increase in methane in the air, because cattle and rice fields release this gas.

Deforestation leads to an increase in carbon dioxide in the atmosphere. The trees may be burned after they are cut down, releasing carbon dioxide. Decomposers break down wood chips and branches, and they release carbon dioxide by respiration. Photosynthesis uses carbon dioxide from the atmosphere. If there are fewer trees to photosynthesise, less carbon dioxide will be absorbed.

Command word: Explain

If you are asked to **explain** something, you need to make it clear, or state the reasons why it happens. You should not just give a list of reasons.

Command word: Describe

If you are asked to **describe** something, you need to recall some facts, events or processes in an accurate way.

This part of the answer gives a general explanation of why deforestation happens.

Three detailed reasons for deforestation are given here. Two of these reasons are linked to an increase in methane in the atmosphere.

Trees and other plants 'lock up' carbon dioxide as carbon compounds in their tissues.

Two main effects of deforestation are described. These are how carbon dioxide is released, and how less carbon dioxide is absorbed. Both lead to an increase in levels of this gas in the atmosphere.

You need to organise your answer into a logical order, with ideas linked in a sensible way. Take care to include only the information required by the question. In this case, although increases in carbon dioxide and methane are a cause of global warming and climate change, there is no need to describe or explain these processes.

Now try this

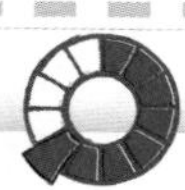

The factory farming of animals is one way to increase the efficiency of food production. Describe the advantages and disadvantages of the factory farming of pigs to provide meat for human consumption. **(6 marks)**

You can revise farming techniques on page 101. Remember that antibiotics may be used in agriculture (revise the problems this causes on page 82).

Answers

Extended response questions

Answers to 6-mark questions are indicated with a star (*).
In your exam, your answers to 6-mark questions will be marked on how well you present and organise your response, not just on the scientific content. Your responses should contain most or all of the points given in the answers below, but you should also make sure that you show how the points link to each other, and structure your response in a clear and logical way.

Paper 1

1. Microscopes and magnification

total magnification = $5 \times 20 = \times 100$ **(1)**

actual length = 15/100 = 0.15 mm **(1)**

2. Animal and plant cells

1 cell membrane **(1)**; nucleus **(1)**; cytoplasm **(1)**; mitochondria **(1)**; ribosomes **(1)**

2 Muscle cells respire more/need more energy **(1)**. Mitochondria are where respiration happens/energy is released **(1)**.

3 Plant roots are in the dark **(1)** so they do not carry out photosynthesis **(1)**.

3. Eukaryotes and prokaryotes

(a) (i) 7.5×10^{-5} m **(1)**

(ii) 7.5×10^{-7} m **(1)**

(b) two orders of magnitude **(1)**

4. Specialised animal cells

Ciliated cells are rectangular/column shaped **(1)** so they can fit together to line the trachea **(1)**.

Sperm cells are streamlined/long and thin **(1)** to help them swim more easily **(1)**.

5. Specialised plant cells

a long thin finger-like part **(1)** that grows between soil particles **(1)**

6. Using a light microscope

neat pencil drawing without unnecessary shading **(1)**, cell wall labelled **(1)**, nucleus labelled **(1)**, cytoplasm labelled **(1)**, chloroplast labelled **(1)**, e.g.

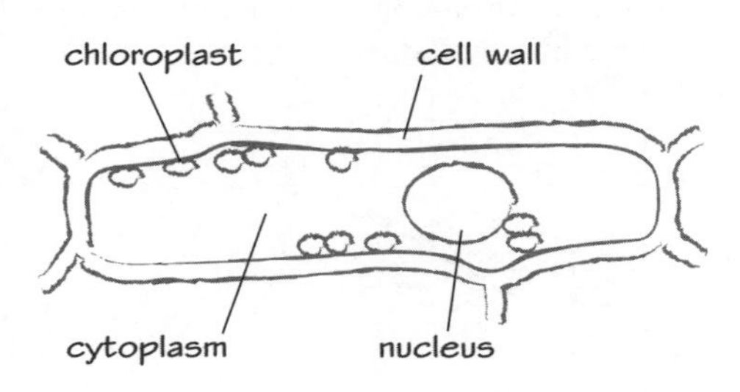

7. Aseptic techniques

1 gamma radiation **(1)**; heating in an autoclave **(1)**; heating in a flame **(1)**

2 This is to kill bacteria on the bench **(1)** that could contaminate the experiment/cultures/to prevent contamination **(1)**.

3 4 hours = 4×60 = 240 minutes, so number of divisions = 240/40 = 6 **(1)** or

40 minutes = 40/60 = 0.667 hours, so number of divisions = 4/0.667 = 6 **(1)**

number of bacteria = $2500 \times 2^6 = 2500 \times 64 = 160000$ **(1)**

8. Investigating microbial cultures

(a) antiseptic B: mean diameter = 25 mm **(1)**; area = 491 mm^2 **(1)**

antiseptic C: mean diameter = 20 mm **(1)**; area = 314 mm^2 **(1)**

(b) Antiseptic B kills the most bacteria **(1)** because it has the largest clear zone **(1)**.

9. Mitosis

DNA is replicated/two copies of each chromosome are produced **(1)**; one set of chromosomes is pulled to each end of the cell (then the nucleus divides) **(1)**.

10. Stem cells

Advantage, e.g. stem cells are genetically identical to the patient's cells/contain the same genes as the patient **(1)**, so the cells will not be rejected by the patient's body **(1)**.

Disadvantage, e.g. egg cell used/embryo created **(1)** and people may have ethical/religious objections to this **(1)**.

11. Diffusion

1 The rate of diffusion increases. **(1)**

2 Diffusion is the spreading out/net movement of particles **(1)** from an area of higher concentration to an area of lower concentration **(1)**.

12. Exchange surfaces

1 having an efficient blood supply **(1)**; keeping the exchange surface ventilated **(1)**

2 Multicellular organisms have smaller surface area to volume ratios **(1)**; exchange surfaces increase the surface area available for exchange of materials **(1)**.

13. Osmosis

1 the diffusion of water **(1)** from a dilute solution to a concentrated solution **(1)** through a partially permeable membrane **(1)**

2 The cell cytoplasm is more concentrated than the water **(1)**, so water moves across the cell membrane into the cytoplasm (causing the cells to expand) **(1)**.

14. Investigating osmosis

(a) 1 mark for each correct percentage change in mass:

Concentration of sucrose (g/dm³)	Change in mass (g)	% change in mass
75	+0.13	+6.0
150	−0.19	−9.3
225	−0.38	−19.4

(b) axes labelled with quantity and unit, and scales chosen so plotted points cover at least 50% **(1)**; all five points plotted ± ½ square **(1)**; curve of best fit is drawn **(1)**; e.g.

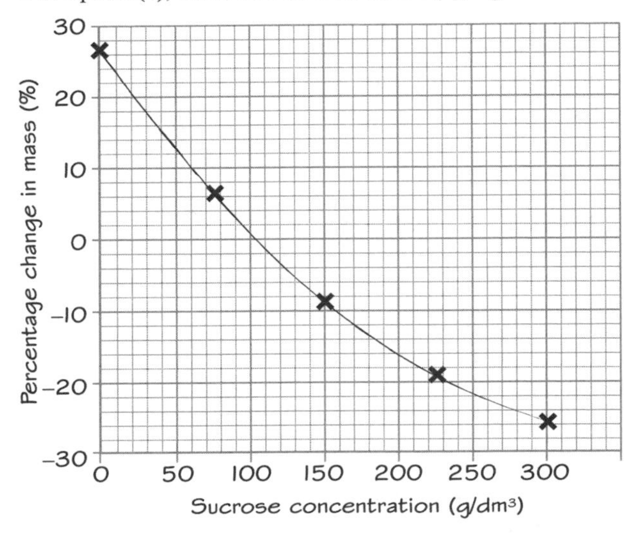

(c) Concentration is estimated from graph at 0% change, e.g. 100 g/dm³. **(1)**

15. Active transport

Diffusion is from high concentration to low concentration, but active transport is from low concentration to high concentration **(1)**; diffusion does not need membranes but active transport does **(1)**; diffusion does not need energy but active transport does **(1)**.

16. Extended response – Cell biology

*Answer could include the following points to 6 marks:

Similarities (both have):

- nucleus
- cytoplasm
- cell membrane
- mitochondria
- ribosomes.

Differences:

- Plant cells have a cell wall.
- Cell wall is made from cellulose.
- Plant cells often have chloroplasts.
- Plant cells often have a permanent vacuole.
- Vacuole is filled with cell sap.
- Plant cells often have regular shapes.

Functions:

- nucleus – contains genes/controls cell activities
- cytoplasm – site of many reactions that take place in the cell
- cell membrane – controls what substances enter or leave the cell
- mitochondria – where respiration happens/release energy for the cell
- ribosomes – site of protein synthesis/where proteins are made
- cell wall – strengthens/supports the cell
- chloroplasts – site of photosynthesis/where photosynthesis happens
- vacuole – helps to keep the cell rigid.

17. The digestive system

suitable summary table, e.g.

	Amylase	Proteases	Lipases
Sites of production	salivary glands pancreas small intestine	stomach pancreas small intestine	pancreas small intestine
Sites of action	mouth small intestine	stomach small intestine	small intestine
Substances broken down	starch	proteins	lipids/fats/oils
Products formed	simple sugars	amino acids	fatty acids and glycerol

1 mark for each correct site of production *and* action for each enzyme, to 3 marks.

1 mark for each correct substance broken down *and* products for each enzyme, to 3 marks.

18. Food testing

It is safer to heat a test tube of liquid using a hot water bath as liquid may boil or escape if a Bunsen burner is used **(1)**; the Sudan III stain is dissolved in ethanol **(1)** which is highly flammable/its vapour could ignite if a Bunsen burner flame is nearby **(1)**.

19. Enzymes

1 pH value given <7 **(1)**; the conditions in the stomach are acidic/stomach produces hydrochloric acid **(1)**

2 The shape of the active site changes/the enzyme is denatured if the temperature becomes too high **(1)** so the substrate no longer fits **(1)**.

20. Investigating enzymes

1 correct values **(1)** given to 2 significant figures **(1)**

pH of buffer	Time taken to digest starch (s)	Rate of reaction (/s)
4	169	0.0059
5	125	1/125 = 0.0080
6	100	1/100 = 0.010
7	118	1/118 = 0.0085
8	260	1/260 = 0.0038

2 axes labelled with quantity and unit, and scales chosen so plotted points cover at least 50% **(1)**; all five points plotted ±½ square **(1)**; curve of best fit drawn **(1)**; e.g.

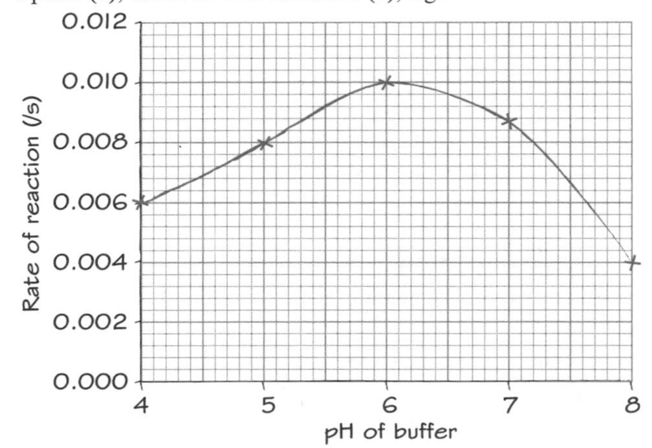

3 The optimum pH is 6 **(1)**; this is when the rate of reaction is greatest/the reaction time is the lowest **(1)**.

4 Temperature also affects enzyme activity/rate of reaction **(1)** so it must be kept the same to make sure the results are valid **(1)**.

21. The blood

Red blood cells are smaller than white blood cells (1); red blood cells do not have a nucleus but white blood cells do (1); red blood cells have a biconcave shape but white blood cells do not (1); red blood cells contain haemoglobin but white blood cells do not (1).

22. Blood vessels

1 artery, capillary, vein (1)

2 Cells need oxygen and glucose for respiration, and to get rid of waste products/carbon dioxide (1); capillaries are the blood vessels that exchange substances with the body cells (1); a short distance between a body cell and a capillary allows efficient/fast diffusion (1).

3 to allow arteries to stretch/not burst when blood is pushed through them (1); to allow arteries to regain their shape afterwards (1)

23. The heart

The right ventricle pumps blood to the lungs (1), but the left ventricle pumps blood to the rest of the body (1).

24. The lungs

Exhaled air has less oxygen (1), more carbon dioxide (1) and more water vapour (1).

25. Cardiovascular disease

Layers of fatty material build up inside (1); this narrows them (1) so less blood can flow through them (1).

26. Health and disease

1 example of a non-communicable disease, e.g. cardiovascular disease/cancer/diabetes (1)

2 any four from: non-communicable disease (1); communicable disease (1); diet (1); stress (1); life situations (1)

27. Lifestyle and disease

unfit men of normal weight/fit men of normal weight (1)

28. Alcohol and smoking

1 brain damage (1); liver damage/cirrhosis (1)

2 Substances in smoke narrow the blood vessels (1), increasing blood pressure (1).

29. The leaf

1 any four from: epidermis/epidermal tissue (1); palisade mesophyll (1) spongy mesophyll (1); mesophyll (1); xylem (1); phloem (1)

30. Transpiration

1 Water enters root hair cell by osmosis (1), moves through plant via xylem (1) and evaporates from leaf as water vapour (1) via stomata (1).

2 More water vapour is lost in the light/during the day (1) than at night/rate of transpiration is lowest at night (1).

31. Investigating transpiration

1 light intensity (1); temperature (1); air movement (1); humidity (1)

2 At higher temperatures, the rate of photosynthesis is greater (1) so more stomata open (1); water molecules move more quickly (1) and when the air moves faster (past the leaves) the concentration gradient of water vapour increases (1), so the rate of diffusion increases (1).

32. Translocation

in the cell sap (1); through pores in the end walls (1)

33. Extended response – Organisation

*Answer could include the following points to 6 marks:

Surface area:

- millions/many alveoli in the lungs
- increases surface area
- larger surface area gives greater rate of diffusion.

Diffusion distance:

- walls are thin
- one cell thick
- provides a short diffusion distance
- shorter diffusion distance gives greater rate of diffusion.

Difference in concentration:

- inhaled air has a high concentration of oxygen/higher concentration than the blood
- large difference in concentration/high concentration gradient gives greater rate of diffusion
- blood carries oxygen away from alveoli
- maintains high concentration gradient
- blood has high concentration of carbon dioxide/higher concentration than inhaled air
- large difference in concentration/high concentration gradient gives greater rate of diffusion
- breathing out carries carbon dioxide away from alveoli
- maintains high concentration gradient.

34. Viral diseases

1 two from the following for 1 mark each: direct contact; by water; by air

2 Viruses exist/live/reproduce in cells (1); burst out/escape from the cell (1).

35. Bacterial diseases

1 They damage tissues/make us feel ill. (1)

2 Washing hands kills/removes bacteria (1), stopping bacteria spreading/getting on the food (1).

36. Fungal and protist diseases

1 Use fungicide to kill the fungi (1); wash contaminated clothing/avoid skin contact with an infected person/avoid contact with contaminated clothing (1).

2 (a) There will be fewer places left for mosquitos to breed. (1)
(b) The nets stop mosquitos reaching/biting people. (1)

37. Human defence systems

1 any two from the following for 1 mark each: skin/scabs; sticky mucus; cilia

2 Lysozyme/hydrochloric acid will not reach pathogens inside larger pieces of food. (1)

38. The immune system

White blood cells ingest pathogens (phagocytosis) (1) and destroy them (1). White blood cells produce antibodies (1) that attach to antigens leading to the destruction of the pathogen (1). White blood cells produce antitoxins (1); these attach to poisonous substances and inactivate them (1).

39. Vaccination

1 Vaccination means giving a vaccine to cause an immune response in the body (1).

2 Infection can only be transmitted from an infected individual; vaccinated poultry are immune and so cannot become infected (1); they will not contain *Salmonella* bacteria so people will not get food poisoning when they eat the poultry (1).

40. Antibiotics and painkillers

1 Measles is caused by a virus **(1)**; antibiotics do not kill/affect viruses **(1)**.

2 one from: viruses exist in the body's cells **(1)**; drugs that kill viruses may also harm the body's cells **(1)**

41. New medicines

1 digitalis/digoxin from foxgloves **(1)** to treat irregular heartbeat **(1)**

OR

aspirin from willow tree bark **(1)** to treat aches and pains/as a painkiller **(1)**

2 The doctors may unintentionally give away clues to the patients (called observer bias) **(1)**; if the patients are able to work out who receives the drug and who receives the placebo, the trial is less reliable **(1)**.

42. Plant disease and defences

1 Cellulose cell walls make it difficult for microorganisms to pass through **(1)**; waxy leaf cuticle protects against microorganisms **(1)**; layers of dead cells around stems/bark make it difficult for pests/pathogens to penetrate **(1)**.

43. Extended response – Infection and response

*Answer could include the following points to 6 marks:

Benefits to vaccinated children:

- Children become immune to measles, mumps and rubella
- usually for life.
- This protects them against the effects of having these diseases (which can be serious).

Drawbacks to vaccinated children:

- possible side effects
- in small numbers of children.

Benefits to the general population:

- Target may be linked to herd immunity.
- Point at which the chance of a non-immunised person coming into a contact with an infected person is very low
- low enough for the pathogen to be lost from the population.
- This protects unvaccinated children and adults.

Other points:

- Benefits greatly outweigh the risks.
- None of the years 1988–2014 met the World Health Organization target
- so benefits vaccinated individuals but not the population as a whole.

44. Photosynthesis

The green parts contain chlorophyll **(1)** which absorbs light/energy for photosynthesis **(1)**; photosynthesis only happens in the green parts **(1)**; starch is made from glucose (in these areas) **(1)**.

45. Investigating photosynthesis

1 axes labelled with quantity and unit, and scales chosen so plotted points cover at least 50% **(1)**; all four points plotted ± ½ square **(1)**; curve of best fit drawn **(1)**; e.g.

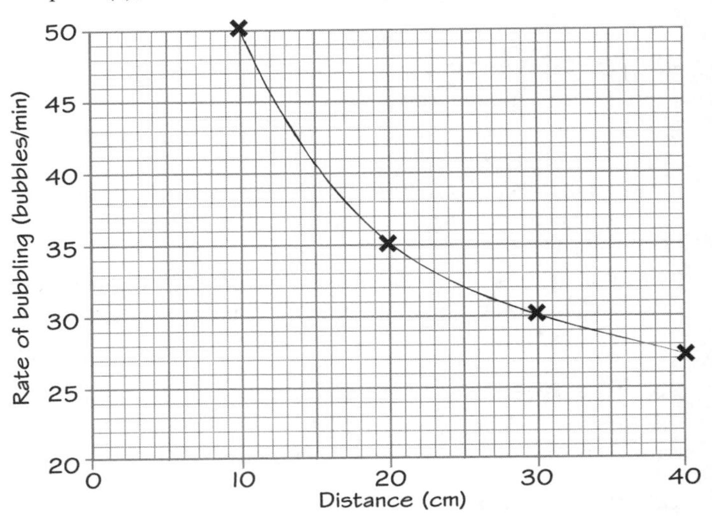

2 As the distance increases the rate of bubbling/photosynthesis decreases **(1)**; the rate of change becomes less at greater distances **(1)**.

46. Respiration

(a) advantage: releases more energy than anaerobic respiration **(1)**; disadvantage: needs oxygen/cannot happen without oxygen/produces carbon dioxide (which must be removed from cells) **(1)**

(b) advantage: does not need oxygen/does not release carbon dioxide **(1)**; disadvantage: releases less energy than aerobic respiration/produces lactic acid **(1)**

47. Responding to exercise

1 Anaerobic respiration occurs/glucose undergoes incomplete oxidation **(1)** and lactic acid builds up in the muscles **(1)**.

2 Heart rate/stroke volume increases **(1)**; breathing rate increases **(1)**; breath volume increases **(1)**.

48. Metabolism

1 (the sum of) all the reactions that happen in a cell or in the body **(1)**

2 amylase/carbohydrase **(1)**; lipase **(1)**; protease **(1)**

49. Extended response – Bioenergetics

*Answer could include the following points to 6 marks:

Increased breathing rate:

- increased muscle contraction
- so more oxygen is needed for respiration
- and more carbon dioxide produced by muscle cells
- rate of gas exchange in the lungs increased
- more oxygen inhaled/diffuses into the blood
- more carbon dioxide exhaled/removed from the body.

Increased heart rate:

- blood is pumped faster around the body/to muscles
- more oxygen delivered to muscles
- more glucose delivered to muscles
- for aerobic respiration
- which releases energy.

Anaerobic respiration:

- reduced need for anaerobic respiration
- so reduces build-up of lactic acid in the muscles.

Paper 2

50. Homeostasis

the regulation of the internal conditions of a cell or organism (1) to maintain optimum conditions for function (1) in response to internal and external changes (1)

51. Neurones and the brain

1 Sensory neurones carry electrical impulses from receptor cells to the central nervous system (1); relay neurones link other neurones together and make up the nervous tissue of the central nervous system (1); motor neurones carry nerve impulses from the central nervous system to effectors (1).

2 The long axon and dendron transmit impulses over long distances (1); dendrites can receive impulses from receptor cells (1); axon terminals can connect with other neurones (1); myelin sheath insulates the neurone from surrounding neurones/helps the electrical impulse to travel faster (1).

52. Reflex actions

Reflex actions do not involve a conscious part of the brain. (1)

53. Investigating reaction times

Practice may improve reaction time (1), so practice before starting would affect the results/reaction times (1).

54. The eye

Circular muscles (of the iris) contract (1); radial muscles (of the iris) relax (1); the pupil constricts/becomes smaller (which lets less light into the eye) (1).

55. Eye defects

1 Near objects are clear with myopia but blurred with hyperopia (1); distant objects are blurred with myopia but clear with hyperopia (1).

2 Hyperopia is caused by the eyeball being too short/the cornea being too flat/ciliary muscles not being strong enough to make the lens thick enough (1), so light focuses behind the retina (1); a convex/converging spectacle lens (1) refracts/converges light (1), so the eye can focus it on the retina (1).

56. Thermoregulation

Blood vessels dilate/vasodilation (1); (more) sweat is produced from the sweat glands (1).

57. Hormones

1 substance secreted by a gland (1); transported in the bloodstream (1); has an effect on a target organ (1)

2 Hormonal responses are slower (1) and have longer-lasting effects (1).

or the opposite for nervous responses.

58. Blood glucose regulation

1 (a) liver (1)

(b) increases absorption of glucose (1); causes glucose to be converted into glycogen (for storage) (1)

2 detects rise/fall/changes in blood glucose concentration (1); secretes insulin (1)

59. Diabetes

More insulin is needed if a big meal is taken rather than a small meal (1); more insulin is needed if a meal contains a lot of sugar/carbohydrates (1); less insulin is needed if the person is exercising (because glucose is used up in respiration) (1).

60. Controlling water balance

from the skin in sweat (1); via the kidneys in urine (1)

61. Kidney treatments

Partially permeable membrane lets small molecules/ions diffuse (1), but not larger molecules/proteins/blood cells/platelets (1).

62. Reproductive hormones

FSH causes an egg to mature (1) but LH causes an egg to be released (1).

63. Contraception

two from the following:

- Use oral contraceptive (1) which contains hormones that inhibit FSH production/stop eggs maturing (1).
- Use injection/implant/skin patch containing progesterone (1) which prevents egg maturation/release (1).
- Use a diaphragm (1) which fits over the cervix/stops sperm entering the uterus (1).
- Use an intrauterine device (IUD) (1) which stops sperm and eggs surviving in the uterus (1).
- Use natural family planning (1) which involves abstaining from intercourse when an egg may be in the oviduct (1).

64. Plant hormones

1 Auxins stimulate cell elongation in shoot cells (1); auxins inhibit cell elongation in root cells (1).

2 Plant shoot will grow towards the light (1); leaves need light for photosynthesis/leaves will get more light for photosynthesis (1).

65. Investigating plant responses

1 to increase the validity of the experiment/results (1); to allow a mean to be calculated (1)

2 The first few millimetres of the seedlings will not be measured (1), so the heights will all be too small (1) by the same length/zero error (1).

66. Extended response – Homeostasis and response

*Answer could include the following points to 6 marks:

- Receptor cells in the eye detect a stimulus.
- The stimulus is the bright light/increase in light intensity.
- Receptor cells pass impulses to sensory neurone.
- Sensory neurone passes impulses to relay neurone.
- Relay neurone is in the spinal cord/central nervous system/CNS.
- Relay neurone passes impulses to motor neurone.
- Motor neurone passes impulses to the effector/muscle in the iris.
- Muscle contracts/makes pupil smaller.
- There are synapses between neurones.
- Chemicals carry impulses across the synapse.

67. Meiosis

Haploid cells have half the number of chromosomes/one set of chromosomes, but diploid cells have the normal number of chromosomes/two sets of chromosomes (1); haploid cells are formed by meiosis but diploid cells are formed by fertilisation (1) and by mitosis of body cells (1); gametes are haploid cells but body cells are diploid cells (1).

68. Sexual and asexual reproduction

1 Strawberry plants produce runners with small plants along them (1), and daffodils produce underground bulbs which divide to form new bulbs (1), by mitosis/asexual reproduction (1).

2 Non-identical gametes (1) are produced by meiosis (1); these join/fuse to form a new individual (1), which involves mixing of the genetic material (1).

69. DNA and the genome

DNA is a polymer made up of repeating nucleotide units **(1)**, made of two strands of phosphate and sugar groups **(1)** in a double helix **(1)**; a base is joined to each sugar **(1)**; the four bases are A, C, G and T **(1)**.

70. Genetic terms

Dominant alleles are expressed if there are two copies of the allele or only one **(1)**; recessive alleles are only expressed if there are two copies of the allele **(1)**.

71. Genetic crosses

(a) Completed Punnett square diagram drawn, e.g.

Gametes	B	B
b	Bb	Bb
b	Bb	Bb

offspring alleles correct **(1)**

(b) There is no chance (0/4, 0, 0%) **(1)** because all the offspring will be heterozygous **(1)** and b/brown is only expressed if there are two copies of the allele **(1)**.

72. Family trees

(a) 10 is tt **(1)** because they are not a taster (and PTC tasting is dominant) **(1)**; 9 is Tt **(1)** because some of his children are not tasters **(1)**.

(b) Individuals 5 and 6 are non-PTC tasters **(1)**; they cannot have the PTC tasting allele **(1)** because this is dominant **(1)**; so their children cannot inherit the PTC tasting allele from them **(1)**.

73. Inheritance

1 Females have the same sex chromosomes/XX chromosomes **(1)**, so egg cells (formed by meiosis) must contain an X chromosome **(1)**.

2 two from the following for 1 mark each:
- Polydactyly is due to a dominant allele, but cystic fibrosis is due to a recessive allele.
- People can have polydactyly if they inherit one or two of these alleles, but people must inherit two of these alleles to have cystic fibrosis.
- People can be carriers of cystic fibrosis/have the allele but do not have the disorder, but there are no carriers of polydactyly.

74. Variation and evolution

1 Identical twins have the same alleles/genotype because they come from the same fertilised egg **(1)**, so characteristics controlled by genes will be the same **(1)**; characteristics affected by the environment may be different between the twins **(1)**.

2 Resistant strains/individuals are not killed by some insecticides **(1)**; they survive to reproduce **(1)** even when the insecticides are used **(1)**.

75. Selective breeding

Select the sheep in the flock that have the most wool **(1)**; breed from these sheep **(1)**; choose sheep with the most wool from the offspring and breed from them **(1)**; repeat this process over many generations **(1)**.

76. Genetic engineering

1 An organism's genome is modified **(1)** to give a desired characteristic **(1)** by introducing a gene from another organism **(1)**.

2 Cut the gene for herbicide resistance from (the genome of) an organism **(1)**; transfer the gene into crop plant cells **(1)**; grow the cells into plants **(1)**.

77. Cloning

embryo transplant, because several embryo cells can be separated for transplant **(1)**, but in adult cell cloning each body cell only produces one embryo **(1)**

78. Darwin and Lamarck

It challenged the idea that God made all the animals/plants/living things **(1)**; there was not enough evidence for it when it was published **(1)**; how inheritance/variation works was not known at the time/until after it was published **(1)**.

79. Speciation

Birds from species in Australia reached New Zealand and so became separated from their original population **(1)**; natural selection favoured different variations in characteristics **(1)**, including flightlessness (as there are no natural predators in New Zealand) **(1)**; these new species are so different from the original species that they cannot interbreed to produce fertile offspring **(1)**.

80. Mendel

Few people read Mendel's work **(1)**; scientists only linked Mendel's unit with genes and chromosomes many years later **(1)**.

81. Fossils

1 Soft-bodied animals do not form fossils as frequently/easily as the bones of vertebrates. **(1)**

2 the remains of organisms from millions of years ago **(1)**, which are found in rocks **(1)**

82. Resistant bacteria

1 two from: antibiotics not prescribed for non-serious/viral infections **(1)**; patients should finish their course of antibiotics **(1)**; restrict agricultural/farming use of antibiotics **(1)**

83. Classification

The chameleon shares more characteristics with the viper, because they are in the same class **(1)**, which is a smaller group of organisms than a phylum **(1)**.

84. Evolutionary trees

(a) Cows and pigs are more similar to each other **(1)** than they are to humans **(1)**.

(b) cow and pig **(1)**, because the common ancestor for cow and hedgehog lived further back in time than the common ancestor for cow and pig **(1)**

85. Extended response – Inheritance, variation and evolution

*Answer could include the following points to 6 marks:

Similarities:
- Both take many generations.
- Both involve selection of certain variants.
- Both involve starting with mixed populations.
- Both involve reproduction using the individuals with a certain characteristic.

Differences:
- Natural selection does not need human intervention but selective breeding does.
- Selective breeding involves aiming for a desired characteristic but natural selection does not.
- Natural selection typically works over many more generations than selective breeding.
- Natural selection can produce new species but selective breeding cannot.

86. Ecosystems

1 organism: a single living individual **(1)**; population: all the organisms of the same species in a habitat **(1)**; community: the populations of all the different species in a habitat **(1)**; ecosystem: the interaction of a community of living organisms with the non-living parts of their environment **(1)**

2 One from: there is very little light under the trees **(1)**; there is not enough light for other plants to photosynthesise **(1)**; there is not enough light/water/room to grow well under the trees **(1)**.

87. Interdependence

two possible results, each with a reason, e.g.

- The population of sparrows would increase **(1)** because sparrowhawks would no longer eat them **(1)**.
- The population of foxes would increase **(1)** because there would be more fieldmice **(1)**.
- The population of rabbits would increase **(1)** because foxes would eat more fieldmice instead **(1)**.
- The population of grass would decrease **(1)** because there would be more sparrows **(1)**.

88. Adaptation

1 The fur is camouflage (against predators) **(1)**, so it needs to be white against the snow in the winter and brown against the ground/plants in the summer **(1)**.

2 two from: small ears **(1)** to reduce heat loss **(1)**; white colour **(1)** for camouflage against snow/ice/to make it more difficult for prey to see them **(1)**; large feet **(1)** for grip on ice/stop bear sinking into snow **(1)**; thick fur/fat **(1)** to reduce heat loss **(1)**

89. Food chains

Predators eat the prey **(1)**; if there are more predators, more prey get eaten **(1)**.

90. Fieldwork techniques

mean number of dandelions in a quadrat = 142/10 = 14.2 **(1)**

area of field = 100 × 200 = 20 000 m^2 **(1)**

estimated population = 14.2 × 20 000/1 = 284 000 **(1)**

91. Field investigations

(a) The number of dandelions increases as the distance from the tree increases. **(1)**

(b) The light intensity also increases as the distance from the tree increases **(1)**; plants need light for photosynthesis **(1)**; if there is more light available, the dandelion plants can make more glucose/can grow better **(1)**.

92. Cycling materials

(a) respiration **(1)**; combustion **(1)**

(b) photosynthesis **(1)**

93. Decomposition

1 Warmth from the sun **(1)** increases the rate of growth of microorganisms so they break down materials faster **(1)**.

2 Air/oxygen is added to the compost when it is turned over **(1)**, increasing the rate of (aerobic) respiration by decay microorganisms **(1)**.

94. Investigating decay

(a) The time taken decreases if the rate of decay increases **(1)**; the rate of reaction increases as the temperature increases **(1)**.

(b) 55 °C is above the optimum temperature of the lipase **(1)**; at high temperatures the lipase/enzyme will be denatured **(1)**, so decay does not happen/the lipase does not break down the milk and the milk does not become acidic **(1)**.

95. Waste management

two from: human population is increasing **(1)**; standard of living is increasing **(1)**; more cars/pesticides/herbicides/fertilisers/waste **(1)**

96. Deforestation

two from: for timber/fuel **(1)**; to clear land for growing rice/crops for biofuel **(1)**; to clear land for cattle **(1)**

97. Global warming

The organisms that cause/transmit tropical diseases can only live where it is warm **(1)**; if the climate gets warmer in other places, these organisms may be able to survive there **(1)** so people in those places may catch the diseases **(1)**.

98. Maintaining biodiversity

three from: breeding programmes **(1)**; protection/regeneration of rare habitats **(1)**; reintroduce field margins/hedgerows **(1)**; reduce deforestation/increase reforestation **(1)**; recycle waste **(1)**

99. Trophic levels

Not all ingested material is absorbed **(1)**; some is egested as faeces **(1)**; some material lost as waste (e.g. carbon dioxide/water in respiration; water/urea in urine) **(1)**; glucose lost in respiration **(1)**.

100. Food security

1 having enough food to feed a population **(1)**

2 three from: increasing birth rate **(1)**; changing diets **(1)**; new pests/pathogens **(1)**; environmental change, e.g. drought **(1)**; costs of production **(1)**; war/conflict **(1)**

101. Farming techniques

Biomass is lost moving from one trophic level to another **(1)**; sardines are at a lower trophic level than tuna **(1)**.

102. Sustainable fisheries

setting fishing quotas **(1)**; closing an area to fishing for several years **(1)**

103. Biotechnology and food

growing *Fusarium* in fermenters **(1)** to produce large amounts of mycoprotein **(1)**; using genetic engineering/modification **(1)** to produce crops with high yields/nutritional value **(1)**

104. Extended response – Ecology

*Answer could include the following points to 6 marks:

Features of factory farming (could be mentioned in advantages or in disadvantages):

- Pigs are kept in small pens.
- The temperature is controlled.
- High-protein foods are used.
- Antibiotics may be given.

Advantages:

- Less energy is transferred from the pigs to the surroundings.
- Pigs need less food.
- High-protein foods encourage faster growth.
- More pigs can be kept in the same area.
- Predators cannot harm/kill the pigs.
- Antibiotics control disease/increase growth.

Disadvantages:

- The pigs' movement is restricted.
- Ethical concerns, e.g. pigs are stressed/overcrowded; people may think it is cruel.
- High-protein foods might have been processed to feed people instead.
- Agricultural use of antibiotics may produce resistant strains of bacteria.
- Heating uses energy/fuel.

Your own notes

Your own notes

Your own notes

Your own notes

Your own notes

Your own notes

Published by Pearson Education Limited, 80 Strand, London, WC2R 0RL.

www.pearsonschoolsandfecolleges.co.uk

Typeset, illustrated and produced by Phoenix Photosetting
Cover illustration by Miriam Sturdee

Content written by Sue Kearsey and Pauline Lowrie is included.

First published 2017

20 19 18 17
10 9 8 7 6 5 4 3 2 1

British Library Cataloguing in Publication Data
A catalogue record for this book is available from the British Library

ISBN 978 1 292 13502 1

Printed in Slovakia by Neografia.

Acknowledgements
The author and publisher would like to thank the following for their kind permission to reproduce their photographs:
123RF.com: kukhunthod 21; **Science Photo Library Ltd:** Herve Conge, ISM 6; **Shutterstock.com:** Buquet Christophe 42, mark higgins 81
All other images © Pearson Education

Text
Page 66 – Revise Edexcel GCSE (9-1) Biology Higher Revision Guide by Pauline Lowrie; Susan Kearsey. Published by Pearson Education Limited © 2016.

Note from the publisher
Pearson has robust editorial processes, including answer and fact checks, to ensure the accuracy of the content in this publication, and every effort is made to ensure this publication is free of errors. We are, however, only human, and occasionally errors do occur. Pearson is not liable for any misunderstandings that arise as a result of errors in this publication, but it is our priority to ensure that the content is accurate. If you spot an error, please do contact us at resourcescorrections@pearson.com so we can make sure it is corrected.